T0328278

Natech Risk Assessment
and Management

Natech Risk Assessment and Management
Reducing the Risk of Natural-Hazard Impact on Hazardous Installations

Elisabeth Krausmann
European Commission, Joint Research Centre
Ispra, Italy

Ana Maria Cruz
Disaster Prevention Research Institute
Kyoto University, Kyoto, Japan

Ernesto Salzano
Department of Civil, Chemical, Environmental
and Materials Engineering, University of Bologna
Bologna, Italy

ELSEVIER

AMSTERDAM • BOSTON • HEIDELBERG • LONDON • NEW YORK • OXFORD • PARIS
SAN DIEGO • SAN FRANCISCO • SINGAPORE • SYDNEY • TOKYO

Elsevier
Radarweg 29, PO Box 211, 1000 AE Amsterdam, Netherlands
The Boulevard, Langford Lane, Kidlington, Oxford OX5 1GB, United Kingdom
50 Hampshire Street, 5th Floor, Cambridge, MA 02139, United States

Notices
Knowledge and best practice in this field are constantly changing. As new research and experience broaden our understanding, changes in research methods, professional practices, or medical treatment may become necessary.

Practitioners and researchers must always rely on their own experience and knowledge in evaluating and using any information, methods, compounds, or experiments described herein. In using such information or methods they should be mindful of their own safety and the safety of others, including parties for whom they have a professional responsibility.

To the fullest extent of the law, neither the Publisher nor the authors, contributors, or editors, assume any liability for any injury and/or damage to persons or property as a matter of products liability, negligence or otherwise, or from any use or operation of any methods, products, instructions, or ideas contained in the material herein.

Library of Congress Cataloging-in-Publication Data
A catalog record for this book is available from the Library of Congress

British Library Cataloguing-in-Publication Data
A catalogue record for this book is available from the British Library

ISBN: 978-0-12-803807-9

For information on all Elsevier publications
visit our website at https://www.elsevier.com/

  Working together
to grow libraries in
developing countries

www.elsevier.com • www.bookaid.org

Publisher: Joe Hayton
Acquisition Editor: Fiona Geraghty
Editorial Project Manager: Maria Convey
Production Project Manager: Julia Haynes
Designer: Matt Limbert

Typeset by Thomson Digital

Contents

v

List of Contributors

G. Antonioni
Department of Civil, Chemical, Environmental, and Materials Engineering, University of Bologna, Bologna, Italy

S. Averous-Monnery
United Nations Environment Programme (UNEP), Division of Technology, Industry and Economics, Paris Cedex, France

A. Basco
AMRA, Analysis and Monitoring of Environmental Risk, Naples, Italy

V. Cozzani
Department of Civil, Chemical, Environmental, and Materials Engineering, University of Bologna, Bologna, Italy

A.M. Cruz
Disaster Prevention Research Institute, Kyoto University, Kyoto, Japan

R. Fendler
German Federal Environment Agency, Dessau-Roßlau, Germany

S. Girgin
European Commission, Joint Research Centre, Ispra, Italy

N. Kato
Department of Naval Architecture and Ocean Engineering, Osaka University, Osaka, Japan

K.-E. Köppke
Consulting Engineers Prof. Dr. Köppke, Bad Oeynhausen, Germany

E. Krausmann
European Commission, Joint Research Centre, Ispra, Italy

G. Lanzano
National Institute of Geophysics and Volcanology (INGV), Milan, Italy

A. Necci
Department of Civil, Chemical, Environmental, and Materials Engineering, University of Bologna, Bologna, Italy

A.M. Pellegrino
Engineering Department, University of Ferrara, Ferrara, Italy

I. Raben
TNO-The Netherlands Organization for Applied Scientific Research, Utrecht, The Netherlands

J. Reinders
TNO-The Netherlands Organization for Applied Scientific Research, Utrecht, The Netherlands

E. Salzano
Department of Civil, Chemical, Environmental, and Materials Engineering, University of Bologna, Bologna, Italy

G. Spadoni
Department of Civil, Chemical, Environmental, and Materials Engineering, University of Bologna, Bologna, Italy

About the Authors

Elisabeth Krausmann *European Commission, Joint Research Centre, Directorate Space, Security and Migration, Italy*

Elisabeth Krausmann is a Principal Scientist with the European Commission's Joint Research Centre (JRC). Having a PhD in Nuclear Physics and Engineering, her research experience includes risk analysis of natural-hazard impact on hazardous installations, nuclear-reactor safety, severe-accident management and consequence analysis. Since 2006 she leads the Natech activity at the JRC which focuses on the development of methodologies and tools for Natech risk analysis and mapping, accident analysis and lessons learning, and capacity building for Natech risk reduction. She is a Steering Group member of the OECD WGCA's Natech project.

Ana Maria Cruz *Disaster Prevention Research Institute, Kyoto University, Japan*

Ana Maria Cruz is a Professor of Disaster Risk Management at Kyoto University. She received a Chemical Engineering degree in 1987, and worked in industry for over 10 years. She later obtained a MSc in Applied Development and a PhD in Environmental Engineering from Tulane University where she pioneered research on Natechs from 1999. She has worked in the private and public sectors, in academia and with government at the local and international levels in four continents. Her research interests include area-wide Natech risk management, risk perception, and protective behavior in communities subject to Natech and climate change impacts. She has published over 40 journal articles and several book chapters, and serves as an Editor for the *Journal of the International Society for Integrated Disaster Risk Management*.

Ernesto Salzano *Department of Civil, Chemical, Environmental, and Materials Engineering, University of Bologna, Italy*

Ernesto Salzano is Associate Professor at the University of Bologna since 2015. His main research activities are in the field of industrial safety. In particular, he studies risks related to the use of flammable substances and risks to critical infrastructures from external hazards, such as Natech risks, domino effects from explosions, and security risks due to intentional acts (terrorist attacks and sabotage). From 1995 to 2015 Prof. Salzano was a researcher at the Institute for Research on Combustion at the Italian National Research Council where he was in charge of the laboratory for studies on substance explosivity and flammability at high pressure and temperature.

Introduction

E. Krausmann*, A.M. Cruz**, E. Salzano†
*European Commission, Joint Research Centre, Ispra, Italy; **Disaster Prevention Research Institute, Kyoto University, Kyoto, Japan; †Department of Civil, Chemical, Environmental, and Materials Engineering, University of Bologna, Bologna, Italy

In Mar. 2011 the whole world watched in shock as a tsunami slammed into a Japanese nuclear power plant, causing a nuclear meltdown and raising the spectre of nuclear contamination with potentially widespread and long-term consequences. Raging fires and explosions at oil refineries in the wake of the massive earthquake that triggered the tsunami also made global headlines. The Cabinet Office of Japan estimated losses in the order of $US 210 billion in direct damage, making the double disasters the most destructive on record.

The past years set a record in the number of natural disasters accompanied by unprecedented damage to industrial facilities and other infrastructures. In addition to the Japan twin disasters in 2011, recent major examples include Hurricane Sandy in 2012 that caused multiple hydrocarbon spills and releases of raw sewage, the damage to industrial parks during the Thai floods in 2011, or Hurricanes Katrina and Rita in 2005 that wreaked havoc on the offshore oil and gas infrastructure in the Gulf of Mexico.

These events clearly demonstrated the potential for natural hazards to trigger fires, explosions, and toxic or radioactive releases at hazardous installations and other infrastructures that process, store, or transport dangerous substances. These technological "secondary effects" caused by natural hazards are also called "Natech" accidents. They are a recurring but often overlooked feature in many natural-disaster situations and have repeatedly had significant and long-term social, environmental, and economic impacts. In the immediate aftermath of a natural disaster, Natech accidents add significantly to the burden of the population already struggling to cope with the effects of the triggering natural event.

Natural hazards can cause multiple and simultaneous releases of hazardous materials over extended areas, damage or destroy safety barriers and systems, and disrupt lifelines often needed for accident prevention and mitigation. These are also the ingredients for cascading disasters. Successfully controlling a Natech accident has often turned out to be a major challenge, if not impossible, where no prior preparedness planning has taken place.

Unfortunately, experience has shown that disaster risk reduction frameworks do not fully address the issue of technological hazards in general, and Natech hazards

Natech Risk Assessment and Management. http://dx.doi.org/10.1016/B978-0-12-803807-9.00001-2

in particular. Also, chemical-accident prevention and preparedness programs often overlook the specific aspects of Natech risk. This is compounded by the likely increase of future Natech risk due to worldwide industrialization, climate change, population growth, and community encroachment in areas subject to these kinds of hazards.

With increasing awareness of Natech risk and a growing body of research into the topic, in some countries first steps have been taken toward implementing stricter regulations for the design and operation of industrial activities in natural-hazard prone areas. Nonetheless, dedicated risk-assessment methodologies are lacking, as is guidance for industry and authorities on how to manage Natech risks in their installations and offsite. It is necessary to revisit existing expertise and practices in risk management at industrial facilities and to implement dedicated measures for Natech risk reduction where gaps are identified.

This book aims to address the entire spectrum of issues pertinent to Natech risk assessment and management in an effort to support the reduction of Natech risks. While in principle also natural-hazard triggered nuclear and radiological accidents qualify as Natech events, the book focuses on Natech risk management in the chemical industry. Nuclear risks are governed by different legislation and mature risk-assessment methodologies that allow the evaluation of natural-hazard impacts are available.

The book is also intended to serve as a reference information repository and state-of-the-art support tool for the industry user, government authorities, disaster risk reduction practitioners, and academia. In its 15 chapters, it collects modeling and assessment approaches, methodologies and tools, and presents measures to prevent Natech accidents or to mitigate their consequences.

Chapters 2–4 are introductory chapters that discuss the characteristics of Natech risk by giving detailed descriptions of selected Natech accident case studies at chemical plants, pipelines, and offshore infrastructure. This is complemented by general and natural-hazard specific lessons learned and a discussion of the treatment of Natech risk in regulatory frameworks for chemical-accident prevention and preparedness currently in force. Chapters 5 and 6 address the prediction and measurement of natural hazards from an engineering point of view and introduce the characteristics of technological hazards arising from the use of hazardous materials. Chapters 7–12 are dedicated to Natech risk assessment. They outline the general assessment process and describe the different approaches available. Examples of existing Natech risk-assessment methodologies are presented and applied to case studies. Chapters 13 and 14 focus on structural and organizational prevention and mitigation measures. These measures range from engineered safety barriers to early-warning systems and risk governance. Finally, Chapter 15 summarizes recommendations for effective Natech risk reduction and remaining gaps that need to be addressed in the future. A glossary of terms is included in the Annex.

Past Natech Events

E. Krausmann*, A.M. Cruz**
*European Commission, Joint Research Centre, Ispra, Italy; **Disaster Prevention Research Institute, Kyoto University, Kyoto, Japan

Numerous past Natech accidents are testimony to the dangers that can arise when the natural and technological worlds collide. Past experience also teaches that Natech accidents can in principle be triggered by any kind of natural hazard and that it does not necessarily require a major natural-hazard event, such as a strong earthquake or a hurricane, to provoke the release of hazardous materials. In fact, Natech accidents were often caused by rain, lightning or freeze, to name a few. This chapter provides selected examples of past Natech events to show the wide variety of natural hazard triggers and also the multitude of hazardous target infrastructures.

2.1 CHARACTERISTICS OF NATECH EVENTS

The extent and consequences of Natech accidents have often reached major proportions, indicating low levels of preparedness. The reasons are manifold and cannot be attributed to a single determining factor. The main difficulty probably stems from the fact that Natech risk reduction is fundamentally a multidisciplinary topic that cuts across traditional professional boundaries. In addition, Natech risk is still considered somewhat of an emerging risk that has only been receiving more attention after a number of recent major accidents. As a consequence, there is still a lack of dedicated knowledge on the dynamics of Natech accidents and not much guidance for industry and authorities on how to address this type of risk. This makes scenario development for planning purposes very difficult.

Another factor is the misconception that engineered and organizational protection measures implemented to prevent and mitigate the so-called conventional industrial accidents would also protect against Natech events. However, Natech accidents differ significantly from those caused by, for example, mechanical failure or human error, and therefore require targeted prevention, preparedness, and response activities.

For instance, when impacting hazardous installations, natural hazards can trigger multiple and simultaneous loss-of-containment (LOC) events over extended areas within a very short timeframe. This was observed repeatedly in areas hit by strong earthquakes where it has posed a severe strain on emergency responders who are usually not trained and equipped to handle a large number of hazardous-materials (hazmat) releases simultaneously. In these situations, the risk of cascading events is high.

Furthermore, the very natural event that damages or destroys industrial buildings and equipment, can also render inoperable engineered safety barriers (e.g., containment dikes, deluge systems) and lifelines (power, water for firefighting or cooling, communication) needed for preventing an accident, mitigating its consequences, and keeping it from escalating. This can complicate the successful containment of hazmat releases.

In case of strong natural disasters with releases of dangerous substances, simultaneous emergency-response efforts are required to cope with the consequences of the natural disaster on the population and the Natech accident that poses a secondary hazard. This undoubtedly leads to a competition for scarce resources, possibly leaving some urgently needed response mechanisms unavailable. The hazmat releases can also hamper response to the natural disaster when toxic substances, fires, or explosions endanger the rescue workers themselves. During some Natech accidents an evacuation order was issued to first responders when their lives were in danger, which sometimes meant leaving behind people who were trapped in residential buildings and in need of help.

Another complicating factor is that civil-protection measures, commonly used to protect the population around a hazardous installation from dangerous-substance releases, may not be available or appropriate in the wake of a natural disaster. For instance, in case of toxic releases during conventional technological accidents, residents in close proximity to a damaged chemical plant would likely be asked to shelter in place and close their windows. This measure would not be applicable after an earthquake as the integrity of the residential structures might be compromised. Similarly, evacuation might prove difficult in case roads have been washed away by a flood or are obstructed by a landslide.

These examples clearly show how the characteristics of Natech events differ significantly from those of conventional technological accidents both in terms of prevention and mitigation. They also highlight why the management of Natech accidents can be challenging without proper planning, and it clarifies why dedicated assessment methodologies and tools are necessary for addressing this type of risk.

2.2 KOCAELI EARTHQUAKE, 1999, TURKEY

The M_w 7.4 Kocaeli earthquake in Turkey occurred at 3 a.m. on Aug. 17, 1999. The earthquake killed over 15,000 people and left over 250,000 homeless. The earthquake affected an area of 41,500 km^2 from west Istanbul to the City of Bolu (Tang, 2000). Modified Mercalli Intensity (MMI) values for the Kocaeli earthquake ranged from VIII to X. Roads, bridges, water distribution, power distribution, telecommunications, and ports were also heavily damaged by the earthquake (Steinberg and Cruz, 2004). The affected area was densely populated and the most industrialized region of the country. Not surprisingly, the earthquake damaged more than 350 industrial firms in the Izmit Bay area and resulted in multiple hazardous-materials releases. Cruz and Steinberg (2005) reported that releases occurred from about 8% of industrial plants subject to MMI≥IX.

Steinberg and Cruz (2004) carried out field visits and interviews in the affected areas about a year after the earthquake. The authors visited and analyzed hazardous-materials releases at 19 industrial facilities impacted by the earthquake. They found that 18 of the facilities they visited suffered structural damage, while 14 reported hazardous-materials releases during or immediately following the earthquake. The authors reported that nine of these facilities suffered severe chemical releases, while an additional five had only "minor" releases with little or no adverse effects. It is important to note that some facilities reported multiple releases.

Steinberg and Cruz (2004) identified over 20 cases of hazardous-materials releases in their study. They found that structural failure as the primary cause of the release was reported in six of the 14 facilities that had hazardous-materials releases. Eleven facilities reported liquid sloshing in storage tanks, while nine of these plants indicated that liquid sloshing was the main cause of hazardous-materials releases at their plant. The authors found that damage to containment dike walls resulted in spills in four of six cases while rupture of pipes and connections led to chemical releases in six out of 12 cases. The most important hazardous-materials releases they reported include:

- The intentional air release of 200,000 kg of anhydrous ammonia at a fertilizer plant to avoid tank overpressurization due to loss of refrigeration.
- The leakage of 6.5 million kg of acrylonitrile. The acrylonitrile was released into air, soil, and water from ruptured tanks and connected pipes at an acrylic fiber plant.
- The spill of 50,000 kg of diesel fuel into Izmit Bay from a broken fuel-loading arm at a petrochemical storage facility.
- The release of 1.2 million kg of cryogenic liquid oxygen caused by structural failure of concrete support columns in two oxygen storage tanks at an industrial gas company.
- The multiple fires in the crude-oil unit, naphtha tank farm, and chemical warehouse, the exposure of 350,000 m^3 of naphtha and crude oil directly to the atmosphere, and the LPG leakages and oil spills at the port terminal at an oil refinery.

In the following sections, two major Natech accidents triggered by the Kocaeli earthquake are described. The accident reports are based on personal communication with refinery officials, municipal fire fighters, and other responders during field visits carried out in 2000 and 2001, and work published by Steinberg and Cruz (2004), Cruz and Steinberg (2005), and Girgin (2011), unless otherwise indicated.

2.2.1 Fires at a Refinery in Izmit Bay

2.2.1.1 Accident Sequence and Emergency Response

The affected refinery was operating at full capacity at 3 a.m. the morning the earthquake occurred. The refinery was subjected to strong ground motion, surface faulting, and a tsunami wave (Tang, 2000; Tsunami Research Group, 1999). The combined earthquake effects resulted in three fires and multiple hazardous-materials releases at the refinery. The first fire occurred at the facility's chemical warehouse. It is believed that chemicals stored on the shelves fell down due to the strong ground motion, breaking glass containers and resulting in the spreading and possible mixing and reacting of the released chemicals. The fire, which was ignited either by sparks or by exothermic chemical reactions, was extinguished in less than half an hour (Girgin, 2011; Steinberg and Cruz, 2004).

The second fire started at the crude-oil processing plant due to the structural failure and collapse of a reinforced concrete stack tower measuring 115 m. The collapsed stack fell over the crude-oil charge heater and a pipe rack, breaking 63 product and utility pipes. The fire, which started when the highly flammable substances found in the pipes ignited, was put out by fire-fighters after 4 h, reignited around noon, was extinguished again, and then reignited again at about 6 p.m. on the same evening of the earthquake. The fire kept reigniting due to the continued supply of fuel from the broken pipes. The number of shut-off valves on the product pipelines was insufficient in the vicinity of the charge heater, and flow from the product lines could not be stopped.

The third fire occurred at the refinery's naphtha tank farm (Fig. 2.1). Four floating roof naphtha storage tanks were simultaneously ignited following the earthquake. The fires were caused by sparking resulting from metal-to-metal contact between the metallic seals and the tank walls due to the bouncing of the floating roofs against the inner side of the tanks. The earthquake also caused damage to a flange connection in one of the tanks which resulted in naphtha leakage into the refinery's internal open-ditch drainage system. The fire on the roof spread to the flange connection and through the drainage system to two more naphtha tanks about 200-m away.

Control of the fires was difficult due to the competing fires ongoing at the refinery. Initially, the fires were brought under control, but this was only temporary. Multiple fire-fighting teams arrived to provide support (e.g., from the military, the municipalities, and neighboring facilities). Nonetheless, the fire-fighting teams lost control of the fire due to the major conflagration and had to retreat. Efforts to control the fire were made by land, sea, and air. The loss of electricity and shortage of foam chemical hampered the response activities. Flyovers to throw seawater and foam over the fires were carried out by forest-fire- and carrier airplanes. However, these efforts were also not effective as it was not possible to fly low enough to approach the fires. Fire-fighting operations were abandoned at around 7 p.m. the day of the earthquake, and an evacuation order was issued by the crisis center for a zone of 5-km around the refinery.

FIGURE 2.1 Snapshots of the Tank Farm Fire After the Flange Failure

From Girgin (2011).

The Turkish government requested international assistance, which arrived on the second and third days. Efforts were concentrated on preventing the fires from spreading to other parts of the plant and nearby chemical installations.

2.2.1.2 Consequences

Direct impacts and cascading events resulted in severe consequences at the refinery, and had repercussions offsite. The collapsed stack heavily damaged the crude-oil processing plant and pipe rack resulting in a large fire that took hours to control. According to Kilic and Sozen (2003), the stack tower collapsed not due to lack of strength caused by design or material deficiencies but due to the presence of reinforcing-bar splices in the region where flexural yielding occurred.

At the naphtha tank farm, six naphtha tanks were completely destroyed burning 30,500 t of fuel. The fires also damaged five additional storage tanks due to fire impingement. Heat radiation burned one of the two wooden cooling towers at the plant. The second cooling water tower collapsed due to the earthquake, also affecting the connected water pipes which in turn slowed down fire-fighting efforts. The

LPG, crude oil, and gasoline storage tanks at the refinery, a large ethylene storage tank at a neighboring facility, and two large refrigerated ammonia gas tanks at a fertilizer plant nearby were unaffected by the fire. However, the ammonia tanks were intentionally vented to avoid overpressurization due to the loss of refrigeration capabilities caused by the earthquake.

All in all, the earthquake damaged a total of thirty storage tanks in the naphtha, crude oil, and LPG tank farms. Typical damage to naphtha and crude-oil tanks included elephant-foot buckling of tank walls, bulging of tank tops due to liquid sloshing, cracking of tank roof-shell wall joints, and damage to roof seals. Damage to roofs resulted in exposure of more than 100,000 m^3 of naphtha, and over 250,000 m^3 of crude oil directly to the atmosphere, increasing the threat of additional fires (Cruz, 2003). Cruz (2003) reported that all the legs of the pressurized LPG steel tanks were twisted severely, resulting in an LPG release from a broken flange connection. There was no ignition of the LPG reported at the refinery. However, two truck drivers were killed in a fire from the gas leak just outside the refinery. The fire was believed to have been triggered by an ignition source from one of the trucks (EERI, 1999).

The earthquake also caused extensive damage to the refinery's port facilities, and onsite utilities. Several large-diameter pipelines located near the shoreline, used to transport crude oil from tankers to the storage tanks, fell from their concrete supports but did not break. The loading and unloading jetty was damaged heavily resulting in an oil spill. Nonetheless, sea pollution was largely attributed to oily water runoff from the fire-fighting efforts (Girgin, 2011).

2.2.1.3 Lessons Learned

The multiple fires in the naphtha tank farms called for a reevaluation of floating-roof systems to ensure that during strong ground shaking no metal-to-metal contact occurs. Liquid sloshing caused the sinking of roofs and extensive damage to tanks. While liquid sloshing cannot be prevented, storage tanks can be reinforced to make certain that they do not suffer deformation due to the lateral forces induced by the sloshing liquid.

The spreading of the fires through the internal drainage canal indicates the need for a system of shut-off doors in internal drainage canals to keep spilled oil from entering public areas such as water bodies, sewer systems, or other parts of the plant.

The inability to control the fire at the crude-oil processing unit stemmed from the fact that there were insufficient shut-off valves to stop the flow of flammable materials through the broken pipes. The installation of emergency shut-off valves in critical segments of pipelines is recommended.

Other major problems observed during the Natech accidents at the refinery concerned failures regarding mainly mitigation and emergency response to the Natech accidents. Girgin (2011) summarized these problems based on work by Kilic (1999) as follows:

- lack of foaming systems at the tanks,
- inadequate power generators,
- inadequate diesel pumps,

- limited application of sprinkler systems,
- noninteroperable fire-fighting water connections,
- insufficient containment ponds,
- lack of fire-fighting towers, and
- deficiencies in the coordination and management of the fire-fighting activities.

Based on the lessons identified from these Natech accidents, Girgin (2011) reports that corrective measures have been taken at the refinery since the earthquake. A revision of the emergency-response plan was made taking Natech events into consideration. In particular, the plan now considers the possibility of multiple accidents simultaneously. In order to secure proper coordination and emergency management, regular meetings with all refinery personnel are now held every 2 months to discuss emergency-response practices based on probable scenarios that include Natech risks. Several improvements were made to ensure that there is adequate fire-fighting water supply (e.g., the water capacity was increased and a seawater connection was introduced to the fire-fighting water system) and adequate fire-fighting equipment (e.g., portable diesel pumps with increased capacity, increased length of water hoses, upgrade of fire-fighting vehicles, fixed and mobile water cannons). Water sprinkler and foaming systems have been installed at all tanks. Other measures that have been taken include the installation of gas and flame sensors for the detection of gas leaks and fires, and the maintaining of a higher length of oil containment booms in the case of spills into the sea. The refinery has used the experience from these Natech accidents to learn and improve their disaster preparedness.

2.2.2 Hazardous-Materials Releases at an Acrylic Fiber Plant

2.2.2.1 Accident Sequence and Emergency Response

At an acrylic fiber plant located in Yalova on the south shore of the Marmara Sea, acrylonitrile (AN), a highly toxic, flammable, and volatile liquid, was stored in eight fixed-roof steel tanks at the time of the earthquake. Of these, three partly full tanks suffered major damage during the earthquake from which AN was released. The releases occurred due to sloshing of the liquid in one partly full tank, buckling of the roof in another tank exposing the chemical to the atmosphere, and from a broken outlet pipe in the third tank. 6.5 million kg of AN were released. Simultaneously, the earthquake cracked the concrete containment dikes around the AN tanks, which allowed the leaked chemical to flow through the containment dikes into the plant's storm-water drainage channel and to the Bay of Izmit.

The emergency response to the AN spill was complicated due to external power outage, damage to the plant-internal power generation turbine, loss of communications, and impassable roads due to debris from the earthquake impact on buildings and roads. Considering that AN is highly volatile and flammable, emergency-power generators strategically placed near the tanks were not started as there was concern that a spark could ignite the AN vapors. Emergency generators and pumps had to be moved to a safe place before they could be put into operation. Fortunately, there were no fires as a result of the AN spill.

Emergency response consisted in applying foaming chemical mixed with water to the release to prevent vaporization. However, the application of foam was delayed due to a loss of water supply as the main water line in the city of Yalova had been damaged by the earthquake. External emergency responders from Yalova and the air force arrived on the same day, but the responders could not enter the spill zone because they did not have appropriate clothing or breathing apparatus to protect them from the harmful effects of the toxic chemical. Eventually, the chemical plant ran out of foaming agent and requested additional supplies from the government and nearby industrial facilities. Additional foam and other resources were brought in by sea or by helicopter. The release was finally contained 48 h after the earthquake (Girgin, 2011; Cruz, 2003).

2.2.2.2 Consequences

Environmental effects were observed as a result of the AN leakage into the air, soil, ground water, and the sea. AN air concentrations were lethal to all animals at a small zoo inside the facility about 200 m from the tanks. Dead vegetation in an area of the same radius was also observed, and domestic animals were reportedly killed in nearby villages (Girgin, 2011; Cruz, 2003). There was also a fish kill reported due to the leakage of AN into Izmit Bay (Türk, 1999).

Poisoning symptoms were reported in the nearby villages and included hoarseness, vertigo, nausea, respiratory problems, skin irritation, headache, eye and nasal irritation (Girgin, 2011; Şenocaklı, 1999). A survey among the residents of Altinkum, which lies about 650 m away from the facility, found that the majority of respondents suffered acute toxicity effects. Less severe effects were reported by residents living as far as 2 and 2.5 km from the facility (Emiroğlu et al., 2000). It was also reported that 27 response workers were poisoned, one member of the military fire-fighting team went into a coma, while others team members were seriously affected (Girgin, 2011).

The toxic release and the subsequent evacuation order given by local authorities hampered search and rescue operations, which were mainly conducted by local people due to a lack of professional search and rescue teams. Similar to the case around the refinery discussed in Section 2.2.1, the evacuation order resulted in local people having to abandon search and rescue of family members, friends, and neighbors.

Soil contamination problems became apparent when harvested products in the affected areas that were put on the market were found to be contaminated with AN. The local government had to issue a ban on the agricultural products of the affected areas (Girgin, 2011).

The groundwater under the tanks reached AN concentrations in the thousands of ppm. After about a year of continuous application of a pump-and-treat regimen, the concentrations had dropped into the hundreds of ppm. There was concern, however, about the long-term effects of AN on the ecosystem of the Bay of Izmit as well as on the affected population.

2.2.2.3 Lessons Learned

The simultaneous damage to three AN storage tanks demonstrated the need for improved storage-tank design to resist the strong ground motion. During the reconstruction, the AN tanks were strengthened against sloshing effects and secondary

FIGURE 2.2 New Foam Spraying System Secured Along a Reinforced Concrete Containment Dike

Photo credit: A.M. Cruz.

roofs were constructed to minimize evaporation in case of a leak. Flexible pipe connections between storage tanks and pipelines were introduced to minimize leakage during earthquakes. Containment dikes were lined with impermeable materials and concrete dikes were reinforced. Fire-fighting foam spraying systems were installed around the perimeter of the containment dikes as shown in Fig. 2.2.

Other important measures adopted based on lessons from the accident included an increase in the capacity of emergency power generators. Furthermore, to avoid a fire or explosion, electrically classified power generators, ventilators, and pumps were installed. The location of fixed as well as mobile equipment was carefully evaluated to maximize emergency response when needed. Emergency- response plans were reviewed and improved to consider Natech scenarios.

2.3 TOHOKU EARTHQUAKE AND TSUNAMI, 2011, JAPAN

On Mar. 11, 2011, an M_w 9.0 undersea megathrust earthquake off the Pacific coast of Tohoku shook large parts of Japan. The Tohoku or Great East Japan earthquake not only affected a large number of hazardous installations, causing the release of hazardous materials, it also triggered a tsunami of unexpected magnitude which led to even more damage and destruction among chemical facilities. A field survey carried out by the Japanese Fire and Disaster Management Agency found damage at 1404 oil-storage or petrochemical plants due to the earthquake, although no details on the number and type of hazmat releases were provided (Nishi, 2012). In a parallel survey, the Japanese Nuclear and Industrial Safety Agency collected information on damage and in some cases hazmat releases at 50 high-pressure gas facilities and 139 cases of damage in other hazardous facilities (Wada and Wakakura, 2011).

Also other types of structures processing or housing dangerous substances were affected by the earthquake. For instance, in Ibaraki Prefecture, a tailings impoundment full of mining waste containing arsenic failed during the earthquake due to liquefaction. As a consequence, 40,000 m³ of toxic waste were released that flowed into nearby fields and a river, and eventually reached the coast (JX Nippon Mining and Metals, 2012). According to newspaper reports, the toxic concentration of the released tailings exceeded the value considered safe 25-fold (Asahi Shibun, 2011).

Numerous hazmat releases occurred during the Tohoku earthquake and tsunami. In some cases several Natech accidents were triggered at the same time in the same chemical plant, which led to multiple and simultaneous releases of dangerous substances. In addition, the tsunami aggravated the impacts of earthquake-triggered toxic or flammable releases by washing them away with the floodwaters and spreading them over wide areas. Nonetheless, considering that the management of the tsunami impacts and the nuclear emergency in Fukushima were the first and foremost priority, chemical releases were secondary unless they posed a concrete threat to the population or emergency responders. As a consequence, only very little information on hazmat releases and their impact is available, with the exception of a few major Natech accidents that are well documented.

In the following sections, two major accidents (one triggered by the Tohoku earthquake, one by the tsunami) will be described in detail. Unless specifically indicated, the information is based on interviews with competent authorities and emergency responders who were on duty at the time of the disaster, and on public information documented in Krausmann and Cruz (2013). Supplementary information was made available by the operator of one of the affected refineries (Cosmo Oil, 2011).

2.3.1 Fires and Explosions at an LPG Storage Tank Farm in Tokyo Bay

2.3.1.1 Accident Sequence and Emergency Response

The LPG storage tank farm where the Natech accident occurred is part of a refinery located on the eastern shore of Tokyo Bay. The refinery, which went into operation in 1963, has a production capacity of 220,000 barrels/day (b/d). In addition, it has a total storage capacity of 2,323,000 kL (crude oil, finished and semiprocessed products, lubricating oil, asphalt, and LPG), as well as 26,400 t of sulfur. In Mar. 2015, the maximum LPG storage volume was 46,900 kL in 21 tanks.

The accident was initiated in LPG Tank No. 364, which at the time of the earthquake was under regulatory inspection. As a consequence, it was filled with water instead of LPG to remove air from inside the tank. The main earthquake shock with a peak ground acceleration of 0.114 g caused several of the diagonal braces supporting the

FIGURE 2.3 The Refinery's Tank 364 That Collapsed During the Earthquake

The buckled legs are clearly visible.

Photo credit: H. Nishi.

tank legs to crack. During the 0.99 g aftershock half an hour later, the legs buckled and the tank collapsed, thereby severing the connected LPG pipes and causing flammable LPG releases (Fig. 2.3). The tank met all earthquake design requirements for the area assuming LPG filling. However, with water being 1.8 times heavier than LPG, the support braces and tank legs could not withstand the additional loading due to the earthquake forces. This situation had not been considered in the tanks' design requirements.

The LPG leaking from the ruptured pipes spread and eventually ignited. With the fires heating up the tank contents, the tank adjacent to Tank 364 suffered a boiling liquid expanding vapor explosion (BLEVE), spreading the fire from tank to tank and eventually throughout the whole LPG tank farm. At least five associated explosions were documented at the refinery, the biggest of which created a fireball of about 600 m in height and diameter (Fig. 2.4). Human error contributed to the disaster as a safety valve on one of the LPG pipes had been manually locked in the open position to prevent it from actuating due to minor air leakages during repair work. Once the released LPG ignited, the valve could not be reached and closed, thereby continuously providing LPG to feed the fire. This exacerbated the accident and made the fires burn out of control. By manually overriding the emergency valve, the company was in violation of safety regulations. In a personal communication,

FIGURE 2.4 Fireball at the LPG Storage Farm With a Diameter of About 600 m

Photo credit: National Research Institute of Fire and Disaster, Japan.

the Chiba Prefecture Fire Department expressed its belief that the accident might have been manageable had the safety valve not been locked open.

Debris impact from the exploding LPG tanks damaged asphalt tanks located adjacent to the LPG storage area, causing asphalt to leak onto the ground and into the ocean. Moreover, burning missile projection, and dispersion and ignition of LPG vapors triggered fires in the adjacent premises of two petrochemical corporations. These secondary accidents released methyl ethyl ketone (MEK) and polypropylene. No supporting information could be found for a newspaper report claiming that the heat impingement from the LPG storage blaze sparked a fire in a warehouse containing depleted uranium.

The fires at the LPG storage tank farm were extinguished only after 10 days when the LPG supply was depleted. The emergency-response teams, which comprised on-site, local, regional, and national fire-fighting teams, worked from both land and sea. However, due to the many release sources it was decided to let the tanks burn until the fuel was exhausted. In addition, first responders sprayed water on the burning tanks to accelerate LPG evaporation. The company's emergency-response plan did not take this type of scenario into account, and neither the company itself nor the Chiba competent authorities were by their own admission prepared for an accident of such magnitude.

2.3.1.2 Consequences

The accident caused six injuries at the refinery, one of which was severe. Three injuries were reported in the facility adjacent to the LPG tank farm where a fire was triggered via domino effect. The fires and explosions forced the evacuation of 1142 residents in the vicinity of the industrial area. Pieces of tank insulation and sheet metal were later found at a distance of over 6 km from the refinery and well inside residential areas. Air-quality monitoring in Ichihara Municipality did not indicate excessive amounts of air pollutants due to the fires at the LPG storage tank farm. Overall, there is very little information on the environmental impact of the event, although some asphalt seems to have entered the sea. However, the company indicated that all asphalt was successfully recovered and they highlight that there is no lasting impact on air, water, or soil from the accident.

The accident resulted in significant damage on-site, destroying all 17 storage tanks (Fig. 2.5). The LPG tank farm had to be completely rebuilt and the refinery

FIGURE 2.5 The Refinery's LPG Tank Farm After the Fires and Explosions Triggered by the Tohoku Earthquake

Note the damaged asphalt tanks in the upper left corner of the image.

returned to full operations only in Jul. 2013, more than 2 years after the accident. The explosions also damaged nearby vehicles and ships, and the shock waves broke windows and damaged shutters and roof shingles in nearby residential areas.

In terms of economic losses, the refinery reported a loss on disaster of $US 72 million (based on the average 2011 Yen → $US exchange rate) for the fiscal year 2010 which ended in Mar. 2011. For fiscal year 2011, the company posted a net deficit of $US 114 million, largely due to the suspension of operations at the refinery and associated alternative supply costs.

2.3.1.3 Lessons Learned

Locking the safety valve in the open position in violation of safety regulations caused the accident to escalate out of control. This highlights the importance of adhering to and monitoring safety systems and measures at installations with a major-accident potential. As a consequence of the accident, the company has abandoned the practice of locking the emergency shut-off valve open and it carries out inspections to ensure that all personnel are aware of applicable laws and regulations. In addition, the refinery has ramped up preparedness activities for large-scale disasters by organizing emergency drills.

Another factor in the chain of fortuitous events that led to the disaster regards the filling of LPG tanks with water during inspections. It is considered good practice to not leave the water in the tanks for more than 2–3 days. However, Tank 364 had been filled with water for 12 days already at the time of the earthquake. The company has also addressed this issue by minimizing the time tanks remain filled with water.

From a technical point of view, the tank braces have shown to be the weak point of the structure during earthquake loading. Following the disaster, the LPG tank braces were reinforced to increase the resistance of tanks to potential future earthquakes.

2.3.2 Fires at a Refinery in the Sendai Port Area

2.3.2.1 Accident Sequence and Emergency Response

The refinery is located in the port area of Sendai and has a production capacity of 145,000 b/d. In Mar. 2011 it was subject to both earthquake and tsunami impact. The refinery automatically shut down at a PGA of 0.25 g. The PGA sensors stopped measuring at 0.45 g although it is believed that the actual PGA on site was higher. The inundation depth at the refinery was between 2.5 and 3.5 m.

Multiple accidents occurred at the refinery at the same time. When the tsunami hit, a tanker truck was in the process of loading hydrocarbons in the western refinery section. The truck toppled over and a pipe broke near the truck. Gasoline was continuously released from the break and eventually ignited from sources unknown. Sparking from the truck's battery or static electricity have been advanced as hypotheses. The blaze destroyed the entire (un)loading facility and also engulfed sulfur, asphalt, and gasoline tanks. One gasoline tank was completely melted, others were partially melted and tilted (Fig. 2.6).

FIGURE 2.6 Burned Tanks at the Sendai Refinery Hit by the Tsunami

Photo credit: C. Scawthorn.

In the same refinery section the tsunami caused multiple pipeline breaks with hydrocarbon releases, as well as many flammable leakages from broken pipe connections, which also ignited. The rail-tank loading area was also hit by the tsunami, with some rail cars being overturned by the water and burned by the fires. A significant portion of the refinery's western section was consumed by the flames.

In another two places in the refinery heavy oil was released (Fig. 2.7). In one case, direct tsunami impact damaged a pipe connected to a tank close to the shore that was used for loading ships, spilling 4400 kL of heavy oil. In the second case, the tsunami waters caused a smaller tank to float. Once the waters receded the tank fell back on the ground, thereby breaking an attached pipe. The spill was aggravated because at the time the tsunami slammed into the refinery, a valve on the tank was open as per normal operating procedure because tank filling was underway. The tank contents continued to be released with a total of 3900 kL spilled.

In the eastern part of the refinery, several atmospheric tank roofs vibrated during the earthquake which caused a number of minor spills that stayed, however, confined to the roofs. In addition, although the processing area was not damaged by the tsunami, the earthquake caused some pipe movement and damage and consequently some minor oil spills.

In the Sendai City area of the refinery, the tsunami caused a tanker ship to crash into the pier which resulted in pipeline damage, but no spill occurred. In the

FIGURE 2.7 Hydrocarbon Releases From Damaged Pipelines at Sendai Refinery due to Tsunami Impact

© Google.

Shiogama City area of the refinery, upon hearing the tsunami alarm the workers disconnected the ships which set sail for the open sea.

While the emergency responders knew that there was a fire at the refinery 5 h after ignition due to smoke coming from the facility, they had no means of accessing the site as the debris created by the tsunami had basically obliterated the access roads. Some flyovers with helicopter provided by the Japanese army were made to assess the extent of the fire. Firefighting on land only started on Mar. 15 when an access road was made. The emergency responders had to bring the fire-fighting equipment into the refinery by hand, as a gasoline pipeline that had been moved inland by the tsunami, blocked the gate and had spilled. The on-site equipment could not be used due to tsunami damage.

The fire fighters had to use heavy breathing apparatus to enter the refinery. A significant amount of asphalt was released, as well as sulfur. Some sulfur went underground and continued to burn there. The fire fighters had to drill holes in the ground to be able to inject water to quench the burning sulfur. Mobile pumps were set up at four locations in the western section to pump up river water. Foam was also used but it kept being blown away by the wind. Once the refinery could be accessed there continued to be concern due to the radiation danger from the nuclear accident at the Fukushima Daiichi nuclear power plant. An order was issued according to which all emergency-response activities at the refinery had to stop if radiation levels reached 0.3 μSv. Fortunately, the maximum measured radiation levels did not exceed 0.12 μSv. The fires at the refinery were extinguished after 5 days.

2.3.2.2 Consequences

At the refinery four people were killed by the tsunami. After ignition of the sulfur and the formation of a toxic gas cloud, the mayor issued an evacuation order for a 2-km radius. This also delayed the intervention of the fire fighters. At a park near the refinery, trees and grass were covered with thick oil from the accident. No detailed information on environmental impacts from the accident is available.

Information on economic losses due to the earthquake and tsunami is limited. The refinery owner reported losses of $US 1.2 billion for the Japanese fiscal year 2010. These losses were due to restoration costs, extinguishment of assets, and running cost. The company estimated another $US 300 million losses due to fixed costs during suspended operations for fiscal year 2011. The refinery returned back to full operations in Mar. 2012.

2.3.2.3 Lessons Learned

The widespread damage and fires in the refinery showed that Natech preparedness levels were low and that emergency-response plans did not adequately consider the consequences of tsunami-triggered Natech scenarios. During reconstruction of the damaged and destroyed refinery parts, the owner addressed some of the critical issues identified during the accident. On the one hand, an artificial hill of 5-m height was raised in the eastern part of the refinery to safeguard emergency-response equipment and teams and from which to coordinate response activities in case of a tsunami. On the other hand, the truck (un)loading facility was moved northeast and farther away from the river for better tsunami protection. This followed the realization that it would be more cost-effective to move equipment or facilities out of harm's way as retrofitting or hardening of the facility would have been difficult to achieve.

From an emergency-response point of view it was noted that the communications flow during the crisis needed to be improved to provide for faster response in situations where telephones are inoperable and access roads to hazardous installations are destroyed or blocked by debris from the tsunami.

2.4 SAN JACINTO RIVER FLOOD, 1994, UNITED STATES

In Oct. 1994, heavy rainfall in the wake of Hurricane Rosa caused serious flooding in the San Jacinto River flood plain, which at the time of the floods was crossed by 69 pipelines operated by 30 different companies. The floods caused significant pipeline damage and spills, which due to their magnitude raised concerns about health impacts and environmental pollution, as well as preparedness levels of operators. The accident description in this section is based on the investigation report of the US National Transportation Safety Board (NTSB, 1996).

2.4.1 Accident Sequence and Emergency Response

During the flooding in the San Jacinto River basin a total of eight hydrocarbon pipelines ruptured, releasing LPG, gasoline, crude oil, diesel fuel, and natural gas. The diameters of the affected pipes ranged from 8 to 40 in. Another 29 pipelines were undermined, and in some cases the length of undermining approached or even exceeded the length of unsupported pipe considered safe for continued operation. Pipe failures and undermining occurred both at river crossings and new channels in the flood plain.

Analyses of the failed pipes showed fatigue cracks from multiple origins that were created when the pipelines were uncovered and undermined. This exposed them to the floodwaters which oscillated and deflected them. The forces acting on the pipes caused the pipe walls to bend and buckle, thereby creating fatigue cracks that continued to grow until the pipes could no longer contain the internal pressure created by the hydrocarbons they transported.

The ruptured pipelines continued spewing hydrocarbons into the fast-flowing floodwaters, and petroleum products pooled in areas where the water flow was slow. Gasoline from a 40-in. pipeline carried downriver on the water surface ignited, causing several explosions and fires that began moving southward.

Operator response to the pipeline failures varied significantly although the failures were very similar. Some operators shut down operations but left valves open and hydrocarbons in the pipes while others closed valves and purged the pipeline of product. Other operators continued operations in spite of several pipeline failures they were aware of. It was reported that 24 mainline valves near the river were inaccessible because they were submerged by the flood. The Coast Guard managed the cleanup from the federal side. This involved the laying of booms downstream of the pipeline ruptures to protect sensitive areas. The booms also deflected the liquids to narrow areas where they were collected with skimmers or vacuum trucks. A portion of the released hydrocarbons were burned in situ, although there is some controversy regarding the decision considering that the water levels had dropped and the liquids could have been removed by mechanical means.

2.4.2 Consequences

In some areas, an evacuation order to residents was issued because of strong fumes from petroleum products in the river. All persons within 9 miles of a failed 40-in. gasoline pipeline were evacuated. Once the situation was considered under control, flood evacuees returning to their houses were advised to stay indoors until the air quality had improved. Overall, the pipeline ruptures and the subsequent fires caused 545 injuries in residential areas primarily due to smoke and vapor inhalation. Two residents suffered burns, one of which serious. It was fortunate that much of the area had already been evacuated due to the flooding prior to the Natech accidents. Two pipeline workers sustained injuries when removing their company's damaged pipe. The petroleum fires also caused significant damage to residential and commercial buildings, as well as to cars and boats.

During the floods almost 36,000 barrels of crude oil and petroleum products were released into the San Jacinto River, and 7 billion cubic feet of natural gas were lost. Spill response costs exceeded $US 7 million while losses due to property damage amounted to about $US 16 million (in 1994 $US).

2.4.3 Lessons Learned

The investigation into the pipeline failures during the flood highlighted a number of critical issues related to pipeline design, siting, and operator preparedness. The US National Transportation Safety Board concluded that the most severe damage or ruptures of pipelines occurred in those areas where the river exhibited maximum stream meandering, where sand-mining operations had taken place, and where the river width was constricted due to human constructions which facilitated riverbed scouring. Furthermore, it was found that the design bases of most pipelines that failed or were undermined had not benefitted from an analysis of the dangers that floods could pose to pipeline integrity. With no regulations, industry standards, or guidance available on how to address the hazards of pipeline siting in flood-prone areas, the Safety Board called for the establishment of standards for pipeline design across flood plains or at river crossings.

The Safety Board also criticized the fact that most pipeline operators continued operations in the San Jacinto River valley without assessing the capability of the pipeline design to resist the flood hazard. Also, only few operators took effective response actions to minimize the potential loss of product. It is believed that the failure of the US Department of Transport's Research and Special Programs Administration (RSPA) to require an operator plan with concrete actions to be implemented in case of floods might have contributed to the severity of the Natech accidents.

Another critical issue identified by the Safety Board concerned the lack of effective operational monitoring to promptly identify the location of ruptures and the absence of remote-controlled or automatic valves. This has led to the release of large volumes of hydrocarbons. The Safety Board called upon the RSPA to establish requirements for valves and leak-detection systems in pipelines.

2.5 HURRICANES KATRINA AND RITA, 2005, UNITED STATES

Hurricanes Katrina and Rita impacted the Gulf of Mexico (GoM) in the United States within a short period of time. Hurricane Katrina made landfall, at the upper end of Category 3 intensity with estimated maximum sustained winds of 110 kt, in Southern Louisiana at 11:10 UTC on Aug. 29, 2005 (Knabb et al., 2006a). Less than a month later, while the region was still trying to recover from the impact of Katrina, Hurricane Rita made landfall as an upper Category 3 storm at 07:40 UTC on Sep. 24, 2005, with an estimated intensity of 100 kt, in extreme southwestern Louisiana (Knabb et al., 2006b). Several levees separating Lake Pontchartrain and the city of New Orleans were breached, resulting in the flooding of about 80% of the city. Hurricane Katrina is responsible directly and indirectly for at least 1833 deaths making it one of the deadliest hurricanes in US history (Knabb et al. 2006b).

Combined, the two hurricanes triggered hundreds of hazardous-materials releases from onshore and offshore oil and gas installations. Sengul et al. (2012) reported over 800 releases in 2005 and 400 releases in 2006 due to the hurricane impacts.

Guidry (2006) notes that the volume of onshore and offshore oil spills from the two hurricanes was 30.2 million liters. The hazardous-materials releases and oil spills represented an additional burden on emergency responders and remaining residents, and affected the supply of fuel needed for emergency-response purposes (Cruz and Krausmann, 2009). Although the storms impacted onshore installations as well as offshore infrastructure, the impacts on the offshore oil and gas industry were unprecedented. Thus, in the next section, a summary of the impacts on the offshore industry is presented.

2.5.1 Accident Sequence and Emergency Response

Over 4,000 offshore platforms and more than 50,000 km of pipeline were exposed to the storms (MMS, 2006). Cruz and Krausmann (2009) identified over 611 releases directly attributed to offshore oil and gas platforms and pipelines due to both hurricanes (reported up to May 2006). Table 2.1 presents a summary of the reported hazardous-materials releases from offshore platforms and pipelines due to Hurricanes Katrina and Rita.

Table 2.1 shows that the releases from offshore infrastructure during Hurricane Katrina doubled those reported for Hurricane Rita, with about 80% of the releases occurring from destroyed or damaged platforms and rigs. The large number of releases from platforms was attributed to higher exposure of platforms to the storm forces including hurricane winds, wave action, and currents (Cruz and Krausmann, 2009). There were, however, over 450 offshore pipeline breaks, indicating that pipelines were vulnerable, too. The common practice to deinventory pipelines in preparation for an approaching hurricane helps explain the lower number of oil and gas spills from pipelines in spite of the many storm-induced breaks. The main types of substances released were crude oil and other types of oil (e.g., fuel, lubricating, hydraulic) during Hurricane Katrina, and crude oil and natural gas during Rita. A small number of releases of NO_x were also reported. However, in a large number of cases the types and quantities of material released were not reported or were unknown. Cruz and Krausmann (2009) found that the number of crude oil and other oil releases during Hurricane Katrina was almost three times higher than during Hurricane Rita, although there were twice as many natural-gas releases attributed to Hurricane Rita.

Table 2.1 Hazmat Releases From Offshore Facilities and Pipelines due to Hurricanes Katrina and Rita

	Hurricane Katrina	Hurricane Rita	Total
Platform	366	162	528
Pipeline	42	40	83
Unknown	0	1	1
Total	408	203	611

Adapted from Cruz and Krausmann (2009).

Damage to offshore platforms and pipelines due to the high winds, wave loading, and strong currents were the main cause of the oil and gas spills (Fig. 2.8). The majority (about 88%) of releases during Hurricane Katrina occurred in the front right quadrant, within a 120-km radius from the storm's center as it approached land. This is the area that typically experiences the highest wind speeds. Storm surge was also the highest in this area according to the US National Weather Service (Simpson and Riehl, 1981). The higher number of hazmat releases during Katrina may be attributed to the higher wind speeds during Katrina as compared to Rita as it approached land.

There were also a large number of oil spills reported in the near-shore, including spills from Meraux, LA (Murphy Oil Corporation) in the metropolitan area of New Orleans, as well as coastal areas at the mouth of the Mississippi River at Empire (Chevron Oil), Pilot Town (Shell Oil), and Cox Bay (Bass Enterprises Production Company) (Pine, 2006).

FIGURE 2.8 Oil Rig Washed Aground by Hurricane Katrina

Photo credit: LtCdr. M. Moran, NOAA Corps, NMAO/OAC.

The emergency response and cleanup following the hurricanes was difficult given the scale of damage induced by the hurricanes offshore but also onshore across Louisiana, Mississippi, and Alabama. The extensive damage and flooding meant that damaged facilities and offshore oil infrastructure was practically inaccessible. Hurricane Katrina in particular rendered many roads impassable for weeks or even months. Furthermore, communication and electrical power systems were damaged, in some areas for several months. This resulted in limited or no access to fuel for emergency- and repair vehicles. Furthermore, there was limited personnel available because employees had evacuated outside of the impacted areas before the storms, or due to the extensive flooding which resulted in no place to provide boarding and lodging for workers, cleanup crews, and first responders. Delays in providing emergency response and cleanup occurred due to overwhelmed support resources, such as boats and other vessels, diving equipment, etc. Cruz and Krausmann (2009) reported that by May 2006 only in about 30% of the cases had some remedial action been taken at the time the releases were reported.

A multiagency (including local, regional, and federal agencies) Incident Command Post (ICP) was set up at the US Coast Guard offices in Baton Rouge. The ICP coordinated and prioritized sites for cleanup in order to minimize public health threats and environmental impacts (ICP, personal communication). Cleanup efforts in the coastal marshes were implemented through the use of boats and barges. Coastal cleanup methods included use of in-situ burn techniques, which involved fewer response resources and had proven to be one of the best environmental removal methods for reducing impacts to the sensitive coastal marsh (Pine, 2006).

2.5.2 Consequences

The hurricanes completely shut down oil and gas production, importation, refining, and distribution in the Gulf of Mexico for days or even weeks. In addition, the two storms together caused extensive damage offshore including destroying or damaging 276 platforms, 24 rigs, and 457 pipelines (Cruz and Krausmann, 2008). There were no reported human casualties from offshore infrastructure due to the storms' impacts. The combined extended downtimes, economic losses from the damaged infrastructure, cost of lost products, and environmental clean up resulted in unprecedented losses to the oil and gas industry. This caused a hike in oil prices around the globe in the weeks that followed the storms.

The environmental consequences of the oil and gas spills are less well understood and were even considered low in the direct aftermath of the storms (Koehler, 2007; Pardue et al., 2005). Leahy (2005), however, warned that only time would be able to tell what the true consequences on the coastal ecosystems, and the fisheries and tourism industries might be. Many of the spills affected the immediate area around the location of the spill, or were washed over the coastal wetlands and marshes. In some cases, spilled oil was dispersed by wave action and storm surge. Pine (2006) reports that both the short- and long-term impacts of these oil releases on ecosystems along the many miles of coastline were and will continue to be examined by the teams from NOAA's Office of Response and Restoration. According to Pine (2006), the teams monitor shallow near-shore and wetland environments in areas impacted by chemical releases in an effort to characterize the magnitude and extent of coastal contamination and ecological effects resulting from this storm.

2.5.3 Lessons Learned

Hurricanes Katrina and Rita together triggered a number of hazardous-materials releases from offshore oil and gas platforms and pipelines that was far greater than that reported for previous storms. A high percentage of hazmat releases occurred not more than 50 km from the coastline where storm-surge values were higher and where most platforms, rigs, and pipelines were destroyed or damaged.

The lower number of releases from pipelines demonstrates the effectiveness of the pipeline deinventory practice prior to the storms. On the other hand, the high number of hazardous-materials releases from platforms was attributed to the destruction or damage of the platform by wind or wave action, which resulted in the discharge of oil and other dangerous substances processed or stored on board. Even in cases where the structural integrity of the platforms was untouched, wave action caused inundation of decks and possibly tipping over of storage tanks and containers holding hazardous materials. A lack of more detailed information on the root causes of failures leading to the releases makes it difficult to make precise recommendations on possible improvements in terms of anchoring mechanisms, flood proofing, etc. Cruz and Krausmann (2009) recommended improvements in damage investigation and reporting in order to facilitate an in-depth analysis of the damage and hazmat release dynamics and hence to prevent the recurrence of future releases.

Even one year later, hazardous-materials releases caused by the storms were still being reported, and only about 30% of the releases had been cleaned up or were under some remedial action. This fact indicates that improvements were needed concerning poststorm damage and release assessment, and prestorm planning for poststorm emergency response and clean up.

2.6 MILFORD HAVEN THUNDERSTORM, 1994, UNITED KINGDOM

The case study descriptions in the previous sections are examples of major Natech accidents that can typically accompany natural disasters. However, also natural hazards commonly not considered a serious threat, such as lightning, rain, or freeze, can trigger Natech events with often severe consequences. The following is an example of a major accident in a refinery that was initiated by a thunderstorm. The accident description is based on the investigation report of the UK Health and Safety Executive (HSE, 1997).

2.6.1 Accident Sequence and Emergency Response

The refinery located in Milford Haven was constructed in 1964 and it is one of the largest refineries in Western Europe. It produces gasoline, diesel fuel, kerosene, liquefied petroleum gas (commercial propane and butane), and petrochemical feedstocks with a throughput capacity of 270,000 b/d, including 220,000 b/d of crude oil and 50,000 b/d of other feedstocks, and a storage capacity of 10.5 million barrels. Over the years, significant upgrades to the installation have allowed the plant to increase its production capacity.

The accident on Jul. 24, 1994 involved the refinery's crude distillation unit (CDU), the fluidized catalytic cracking unit (FCCU), and its flare system. The crude oil was separated in the CDU by fractional distillation into naphtha and gas, kerosene, light and heavy diesel, and vacuum gas oil (VGO). The VGO provided the feed for the vacuum distillation unit which in turn fed the FCCU.

The first accident sequence was initiated when a thunderstorm passed over the refinery between 7:20 and 9:00 a.m. Lightning strikes caused a 0.4 s power loss and subsequent power dips throughout the refinery. This caused the repeated tripping of pumps and coolers and consequently led to the lifting of the crude column pressure safety valves. As a consequence, flammable vapors were released and ignited by a lightning strike. The CDU was shut down as a result of the fire.

The power dips caused by the thunderstorm also initiated the second accident sequence which some 5 h later resulted in a powerful explosion and fires. The HSE investigation concluded that a combination of failures in management, equipment, and control systems led to the actual release of hazardous substances and eventually to the explosion. Due to process upsets caused by the power dips, the FCCU briefly lost and then regained VGO feed between 7:47 and 8:00 a.m., leading to feed level fluctuations. The process upset was exacerbated by additional power interruptions at 8:27 and 8:29 a.m. This, together with valve problems that were not recognized, eventually caused the FCCU's wet gas compressor to shut down, resulting in a large vapor load on the FCC's flare system. This led to a high liquid level in the flare knockout drum, thereby exceeding its design capacity and forcing the liquid hydrocarbon into the drum's outlet line which was, however, not designed to take liquid and ruptured due to mechanical shock at an elbow bend (Fig. 2.9). As a result of the rupture at the flare drum's outlet line a pulsing leak appeared, releasing about 20 t of flammable hydrocarbons. The released hydrocarbon liquid and vapor mixture reached explosive levels and drifted inside the process area. The hydrocarbon cloud was eventually ignited by a heater and an explosion occurred 30 s after the outlet line break at a distance of about 110 m from the flare drum. After the explosion, several isolated fires continued to burn within the process area, most importantly a major fire at the flare drum outlet itself.

In response to the accident, the fires were contained and an escalation of the accident was prevented by cooling the nearby vessels containing flammable substances. The explosion and fires incapacitated two of the plant's three flare systems, and consequently plant personnel shut down and isolated the process equipment. Considering that the explosion had disabled the flare relief system, the fires were allowed to burn until the hydrocarbons were exhausted 2.5 days after the explosion.

FIGURE 2.9 The Failed Elbow Bend at the 30-in. Flare Drum Outline Line From Which the Hydrocarbons Were Released

From HSE (1997). © Crown Copyright

The accident was mitigated by the on-site and county fire-fighting teams. After much deliberation, only the on-site emergency plan was activated although it did not consider the possibility of a fire burning for more than 24 h.

2.6.2 Consequences

The accident resulted in 26 nonserious injuries on site. According to the accident investigation report, a disaster was averted because the accident occurred on a weekend when less employees were on site, and because the alkylation unit adjacent to the FCCU withstood the explosion unscathed. This is testimony to the high safety standards that the unit was built and operated to.

Large areas of the refinery suffered severe structural damage due to the explosion and the fires on-site. The blast from the explosion caused damage to buildings, vessels, columns, tanks, pipework, and pipe racks (Fig. 2.10). Block wall buildings

FIGURE 2.10 **Major Damage on Site Caused by the Explosion**

From HSE (1997). © Crown Copyright

near the blast location were completely destroyed. Interestingly, the control room suffered some internal damage because the door had been open at the time of the explosion. Staff had opened the door as the earlier power interruptions had disabled the air conditioning control. The accident did not cause any community disruption as off-site damage was limited due to the refinery's location away from population centers. In Milford Haven town, located at about 3 km from the refinery, some properties sustained glass damage.

As a result of the accident, approximately 10% of the United Kingdom's refining capacity in 1994 was lost during the refinery's downtime (4.5 months). According to Marsh (2003) the monetary losses due to business interruption amount to US$ 70,500,000. Costs related to property damage, debris removal, and cleanup were estimated at US$ 77,500,000 (both numbers refer to 1994 monetary value).

2.6.3 Lessons Learned

While the initiating event of the accident was the process upset caused by the electrical storm, the direct causes of and contributing factors to the explosion were several:

- In the FCCU the debutanizer valve was stuck closed, a fact unknown to the operators. This led to hydrocarbon liquid being continuously pumped into a process vessel that had its outlet closed. Consequently, once the vessel was full, the hydrocarbons entered the pressure relief system and the flare line.
- The operators were overwhelmed by too many alarms triggered by the process upset, and the display screen configuration of the operator control system made it difficult to identify the cause of the incident. It was concluded that the operators were not adequately trained to handle a sustained process upset.
- Tests on instruments whose incorrect behavior contributed to the accident revealed maintenance issues. In addition, the flare drum's outline line was known to be corroded and modifications on the drum's pump-out system resulted in a reduced liquid handling capacity. However, no recorded safety assessment of this modification was available.

Based on these insights, the HSE formulated recommendations for future accident prevention and mitigation in those areas where deficiencies were identified. Most importantly, these concerned a formal and controlled Hazard and Operability (HAZOP) Study in case of modifications; training for staff to better handle unplanned events and perform well under high-stress conditions; and reconfiguration of alarms to facilitate the distinction between safety-critical and other operational alarms. In view of the possibility of prolonged fires, the HSE also recommended considering the availability of adequate supplies of fire-fighting water.

References

Asahi Shibun, 2011. Tohoku tsunami washed arsenic ashore, September 3, 2011. http://ajw.asahi.com/article/0311disaster/quake_tsunami/AJ2011100713655

Cosmo Oil, 2011. Overview of the fires and explosion at Chiba refinery, the cause of the accident and the action plan to prevent recurrence, Press Release August 2, 2011. http://www.cosmo-oil.co.jp/eng/press/110802/index.html

Cruz, A.M., 2003. Joint natural and technological disasters: assessment of natural disaster impact on industrial facilities in highly urbanized areas. Dissertation. UMI Number: 3116886, 204 pp.

Cruz, A.M., Krausmann, E., 2008. Damage to offshore oil and gas facilities following Hurricanes Katrina and Rita: an overview. J. Loss Prev. Process Ind. 21 (6), 620.

Cruz, A.M., Krausmann, E., 2009. Hazardous-materials releases from offshore oil and gas facilities and emergency response following hurricanes Katrina and Rita. J. Loss Prev. Process Ind. 22 (1), 59.

Cruz, A.M., Steinberg, L.J., 2005. Industry preparedness for earthquakes and earthquake-triggered hazmat accidents during the kocaeli earthquake in 1999: a survey. Earthquake Spectra 21 (2), 285.

EERI, 1999. The Izmit (Kocaeli), Turkey earthquake of August 17, 1999. Special Earthquake report, Earthquake Engineering Research Institute, Oakland, CA.

Emiroğlu, C., Koşar, L., Karadağ, K., Abbasoğlu, S., Başçıl, H.S., 2000. AKSA reality. Turk. J. Occup. Health Saf. 3, 12.

Girgin, S., 2011. The Natech events during the 17 August 1999 Kocaeli earthquake: aftermath and lessons learned. Nat. Hazards Earth Syst. Sci. 11, 1129.

Guidry, R.J., 2006. From the incident command center oil spills from hurricanes Katrina and Rita. Proceedings of the Environment Canada Arctic and Marine Oil Spill Program Technical Seminar (AMOP), vol. 2: 989, Vancouver, BC, Canada, June 6–8, 2006.

HSE, 1997. The Explosion and Fires at the Texaco Refinery, Milford Haven, 24 July 1994. Health and Safety Executive, United Kingdom.

JX Nippon Mining and Metals, 2012. Casualties caused by the Great East Japan earthquake and responses at the Oya Mine. http://www.nmm.jx-group.co.jp/english/sustainability/environmental_activities/abolition.html

Kilic, A., 1999. Tüpraş Yangını (TUPRAS Fire). Istanbul Technical University, Istanbul, pp. 7 (in Turkish).

Kilic, S.A., Sozen, M.A., 2003. Evaluation of effect of August 17, 1999, Marmara Earthquake on two tall reinforced concrete chimneys. Struct. J. 100 (3), 357.

Knabb, R.D., Rhome, J.R., Brown, D.P., 2006a. Tropical Cyclone Report, Hurricane Katrina, August 23–30, 2005. National Hurricane Center, United States. http://www.nhc.noaa.gov/data/tcr/AL122005_Katrina.pdf

Knabb, R.D., Brown, D.P., Rhome, J.R., 2006b. Tropical Cyclone Report, Hurricane Rita September 18–26, 2005, National Hurricane Center, United States. http://www.nhc.noaa.gov/data/tcr/AL182005_Rita.pdf

Koehler, A.N., 2007. Acute releases of hazardous substances related to Hurricanes Katrina and Rita. Louisiana Hazardous Substances Emergency Events Surveillance (LaHSEES) System, Louisiana Department of Health and Hospitals, Office of Public Health, Section of Environmental Epidemiology and Toxicology.

Krausmann, E., Cruz, A.M., 2013. Impact of the 11 March 2011, Great East Japan earthquake and tsunami on the chemical industry. Nat. Hazards 67, 811.

Leahy, S., 2005. Years Before Katrina's Environmental Costs can be Measured. Inter Press Service, Rome.

Marsh, 2003. The 100 Largest Losses 1972–2001—Large Property Damage Losses in the Hydrocarbon-Chemical Industries, twentieth ed. Marsh Risk Consulting, New York.

MMS, 2006. Impact assessment of offshore facilities from Hurricanes Katrina and Rita, News Release 3418, January 19, US Minerals Management Service.

Nishi, H., 2012. Damage on hazardous materials facilities, In: Proceedings of the International Symposium on Engineering Lessons Learned from the 2011 Great East Japan Earthquake, Tokyo, Japan, March 1–4, 2012.

NTSB, 1996. Evaluation of Pipelines Failures During Flooding and of Spill Response Actions, San Jacinto River Near Houston, Texas, October 1994, Pipeline Special Investigation Report. National Transportation Safety Board, Washington, DC.

Pardue, J.H., Moe, W.M., McInnis, D., Thibodeaux, L.J., Valsaraj, K.T., Maciasz, E., van Heerden, I., Korevec, N., Yuan, Q.Z., 2005. Chemical and microbiological parameters in New Orleans floodwater following Hurricane Katrina. Environ. Sci. Technol. 39 (22), 8591.

Pine, J.C., 2006. Hurricane Katrina and oil spills: impact on coastal and ocean environments. Oceanography 19 (2), 37.

Sengul, H., Santella, N., Steinberg, L.J., Cruz, A.M., 2012. Analysis of hazardous material releases due to natural hazards in the United States. Disasters 36 (4), 723.

Şenocaklı, M., 1999. İnsanları enkaz altında canlı bırakıp kaçtık! (We escaped leaving people behind under the debris), Vatan, July 10 (in Turkish).

Simpson, R.H., Riehl, H., 1981. The Hurricane and its Impact. Louisiana State University Press, Baton Rouge.

Steinberg, L.J., Cruz, A.M., 2004. When natural and technological disasters collide: lessons from the Turkey earthquake of August 17, 1999. Nat. Hazards Rev. 5 (3), 121.

Tang, A.K., 2000. Izmit (Kocaeli), Turkey, Earthquake of August 17, 1999, including Duzce Earthquake of November 12, 1999. Lifeline Performance, Technical Council on Lifeline Earthquake Engineering (TCLEE), Monograph No. 17, ASCE, Virginia.

Tsunami Research Group, 1999. Izmit Bay Tsunami Survey, August 22–26, 1999, Tsunami Research Group. University of Southern California, Los Angeles.

Türk, E., 1999. Aksa'nın zehiri bölgedeki balıkları öldürdü (Poison of AKSA killed fish in the area), Milliyet, September 9, p. 11 (in Turkish).

Wada, Y., Wakakura, M., 2011. Japan Report, 21st Meeting of the OECD Working Group on Chemical Accidents, October 5–7, 2011, Paris, France.

Lessons Learned From Natech Events

E. Krausmann*, E. Salzano**

*European Commission, Joint Research Centre, Ispra, Italy; **Department of Civil, Chemical, Environmental, and Materials Engineering, University of Bologna, Bologna, Italy

Efforts have been launched to systematically collect and analyze information on the causes and dynamics of Natech accidents, as well as of near misses, to support scenario development and the design of better risk-mitigation options. Using postaccident analysis, conclusions can be drawn on the most common damage and failure modes and hazmat release paths, particularly vulnerable storage and process equipment types, and the hazardous materials most commonly involved in these types of accidents. This chapter gives an overview of natural-hazard specific lessons learned and also discusses features common to Natech accidents triggered by different natural hazards.

3.1 DATA SOURCES AND QUALITY

Lessons can be learned in all phases of risk and accident management, from prevention and preparedness to response and recovery. Depending on the scope of the study, there are analyses of single accidents, which produce immediate lessons specific to the event, or analyses of a set of similar accidents from a broader data pool which yield lessons learned that are more widely applicable. The latter type of study facilitates, for example, the identification of commonly occurring causes of accidents involving specific substances or industries, which may not be easily recognizable within a single occurrence. This analysis also lends itself to identifying technical and organizational risk-reduction measures that require improvement or are missing.

Industrial-accident databases are commonly used for retrieving sets of Natech accident case histories for further analysis. These databases contain accident data from the open literature, government authorities, or in-company sources. Examples of such data repositories are the French ARIA database, the European Commission's Major Accident Reporting System (MARS), The Accident Database of the UK Institution of Chemical Engineers, or the US Coast Guard National Response Center (NRC) database. The quality of information reported in the various industrial-accident databases is not uniform and exhibits different levels of detail and accuracy. This is due to the difficulty of finding qualified information sources, especially in situations where accident reporting by the industry or by authorities is not compulsory, for example, when spill quantities are below reporting thresholds. Data collection has to then rely on voluntary record keeping which is often done by nonexperts. Industrial-accident databases also suffer from a lack of information on near misses, which would be of particular value for learning lessons as they give examples of effective risk-reduction approaches and techniques.

The level of detail is particularly nonuniform for Natech accident data depending on whether the consequences of the Natech event were major or minor, and whether comprehensive information was available for reporting. In addition to the reporting bias toward high-consequence events, industrial-accident databases frequently lack information on the severity of the triggering natural hazard, as well as on failure modes that led to the hazmat release. This makes it difficult to reconstruct the dynamics of the accident and renders the development of equipment vulnerability models linking the natural-hazard severity to the observed damage almost impossible. Consequently, the European Commission has set up the eNATECH database for the systematic collection of Natech accident data and near misses. The database exhibits the more sophisticated accident representation required to capture the characteristics of Natech events and is publicly accessible at http://enatech.jrc.ec.europa.eu.

As yet, most Natech accident analyses concerned accidents triggered by earthquakes, floods, or lightning. Priority was given to these hazards due to the generally greater severity of Natech events caused by earthquakes (Antonioni et al., 2009), and the high frequency of accidents initiated by floods and lightning in the European Union and the OECD (Krausmann and Baranzini, 2012). Systematic analyses of the dynamics and consequences of Natech accidents caused by other natural hazards are scarce.

3.2 GENERAL LESSONS LEARNED

Postaccident analyses of a multitude of Natech events caused by specific types of natural hazards soon revealed that there are certain commonalities regardless of the natural-hazard trigger. These studies indicated, for instance, that atmospheric storage tanks, and in particular those with floating roofs, appear to be particularly vulnerable to earthquake, flood, and lightning impact (Krausmann et al., 2011). While no systematic studies for other types of natural hazards are readily available, individual case histories seem to support this conclusion also in case of storms or heavy rain (MAHB, 2014; Bailey and Levitan, 2008; Godoy, 2007).

From an industrial-safety point of view, the high susceptibility of storage tanks to natural-hazard impact is problematic, as these plant units often contain crude oil, gasoline, or other types of flammable liquid hydrocarbons in large quantities. It is therefore not surprising that many Natech accidents involve hydrocarbon releases that often ignite and escalate into major fires or explosions (Table 3.1). With hazmat

Table 3.1 Substances Involved in a Sample of Flood-Triggered Natech Accidents According to an Analysis by Cozzani et al. (2010)

Substance Category	Number of Accidents
Oil, diesel fuel, gasoline; liquid hydrocarbons	158
Propane, butane, LPG	12
Fertilizers	11
Acid products	7
Cyanides	5
Oxides	5
Ammonia	5
Chlorine	3
Explosives	3
Calcium carbide	3
Soap and detergents	1

releases possibly occurring from several sources at the same time, an increased ignition probability coupled with concomitant damage to safety barriers and systems and the frequent loss of lifelines, the likelihood of domino or cascading disasters is also higher for Natech events compared to conventional industrial accidents.

The good news is that risk mitigation generally seems to pay off. Facilities fare better during natural events if they have benefitted from natural-hazard specific design and the implementation of Natech risk-reduction measures (e.g., Cruz et al., 2016; Pawirokromo, 2014; Cruz and Steinberg, 2005; Bureau and Kokkas, 1992; Lopez et al., 1992). Where these measures are inadequate or totally lacking, damage is more severe or even catastrophic. Problem areas that stand out in most Natech accidents are related to insufficient prevention and preparedness, often caused by the lack of structural design features to withstand the natural-hazard loads, the absence of or the weak enforcement of safety regulations, and by a lack of guidance on how to address the problem of Natech risks in the chemical-process industry.

The biggest challenge for the industry and authorities is a change in mindset to accept that Natech hazards can have significant impacts. In-depth analyses of accident data often indicate grossly inadequate design bases of hazardous installations in natural-hazard prone areas due to the use of generic design criteria instead of acknowledging the specific requirements of process equipment under natural-event loading. Ignorance of the need for Natech-specific additional safety measures and a lack of Natech risk assessment contribute to low preparedness levels. Another common risk factor that was identified is the reliance of industry on external lifelines and emergency-response resources for managing a Natech accident rather than preparing a "stand-alone" emergency plan that considers the failure of response systems during a natural disaster. If response resources are overwhelmed the accident can quickly escalate.

3.3 EARTHQUAKES

The postaccident analysis of data sets related to earthquake-triggered Natech events indicates that multiple and simultaneous hazmat releases are particularly common during earthquakes accompanied by an increased risk of cascading events. Damage to industrial facilities due to direct shaking impact or ground deformation induced by soil liquefaction are the main damage and failure modes of structures built in susceptible areas (e.g., Lanzano et al., 2014; Krausmann et al., 2011). From a safety point of view, damage to buildings or equipment that do not contain hazardous materials is of no immediate concern although the associated economic losses can be huge. The predominant damage modes in this category include elephant-foot- or diamond buckling (Fig. 3.1), the stretching or detachment of anchor bolts causing lateral displacement or uplifting of equipment, or the deformation or failure of support columns and other types of foundation structures. The main hazmat-release mechanisms during earthquakes are tank damage with losses from the tank's roof top due to liquid sloshing, the sinking of floating tank roofs, failure of flanges or rigid tank-pipe connections due to direct shaking impact, or liquefaction-induced ground deformation that leads to pipe ruptures and foundation failures (Lanzano et al., 2015; Krausmann and Cruz, 2013; Zama et al., 2008, 2012; Krausmann et al., 2010). Examples are

FIGURE 3.1 Diamond Buckling of Silos Near the Epicenter of the 2011 Tohoku Earthquake in Japan

Photo credit: E. Krausmann.

FIGURE 3.2 Collapse of a Dryer and Pipe Severing at a Fertilizer Factory Hit by the 2008 Wenchuan Earthquake in China

Photo credit: E. Krausmann.

shown in Figs. 3.2 and 3.3. These releases can be minor or severe. Major releases are caused by tank overturning or collapse due to earthquake loading. Liquid sloshing, for instance, can compromise the structural integrity of tanks which are full or nearly full (Zama et al., 2008, 2012; Salzano et al., 2009; Campedel, 2008).

Release modes from chemical containers in warehouses involve the collapse or overturning of storage racks or the toppling and falling of chemical drums or intermediate bulk containers (IBCs). Dynamics analyses support the conclusion that racks loaded with IBCs are more vulnerable to ground motion than those loaded with drums. The analysis also showed that while robust anchorage reduces the likelihood of rack collapse, it also increases the potential for drum or IBC toppling (Arcidiacono et al., 2014).

Accident analyses also showed that flammable hazmat releases are likely to ignite during earthquakes. Floating roof tanks in particular are prone to fire scenarios as liquid sloshing can bounce the tank's metallic roof against the tank wall, which can create sparks and ignite the tank content. From the statistical analysis of the available accident data set, an ignition probability of 0.76 was calculated by comparing the number of accidents with only release to those with release and ignition (Campedel, 2008). With a reporting bias toward high-consequence events, this number should be considered an upper limit.

FIGURE 3.3 Flange Failure at a Fertilizer Plant in the Area Hit by the Wenchuan Earthquake

Photo credit: E. Krausmann.

3.4 TSUNAMI

A significant number of industrial installations are located in tsunami-prone areas all over the globe. Nevertheless, quantitative data on tsunami damage to industrial structures is scarce, in particular when compared to damage data related to earthquake impacts. The reason is likely related to the low frequency of large tsunami events, and the very recent (post 2004 Indian Ocean tsunami) efforts to start collecting and analyzing information on industry-specific tsunami impacts.

The tsunami triggered by the 2011 Great East Japan earthquake and the extensive damage and destruction it caused in vast parts of the Japanese industry offers an opportunity to learn important lessons. Post-tsunami analyses identified direct hydrostatic and hydrodynamic forces from water inundation, as well as impact forces from water-borne debris as important damage and hazmat-release mechanisms. The high speeds obtained by the water can also drag large floating objects, such as ships, into a hazardous installation which can significantly aggravate the damage severity. Salzano and Basco (2015) contend that while debris impact can be assessed using impact analysis, the large uncertainties related to the type and number of objects (including ships) transported by the tsunami waters render the assessment difficult.

The tsunami forces caused washing away of equipment foundations, tank and pipe floating and displacement, tank overturning and collapse (Fig. 3.4), and the breaking of pipe connections and ripping off of valves which were often accompanied by the release of significant amounts of hydrocarbons (Krausmann and Cruz, 2013; Nishi, 2012). The sheer force of a strong tsunami would be able to affect a large variety of structures and equipment, however, storage tanks, and in particular atmospheric ones, appear to be especially vulnerable.

FIGURE 3.4 Destroyed Heavy Oil Tank at a Thermal Power Plant Battered by the Mega Tsunami in the Wake of the 2011 Tohoku Earthquake in Japan

Photo credit: A. Kouchiyama.

Damage to industrial facilities was also observed in cases where the tsunami compromised the structural integrity of buildings which then collapsed onto the hazardous equipment housed within. FEMA (2012) attributes structural tsunami damage also to wind forces induced by wave motion and scour and slope/foundation failure.

Like earthquakes, strong tsunamis can have a large impact zone, and multiple releases of different types of hazardous substances are expected to occur at the same time with potentially manifold consequences. In addition to possibly driving debris-laden water into a hazardous installation, tsunamis would also widely disperse flammable spills or toxic releases triggered by a preceding earthquake or by the tsunami itself. The ignition probability is high under these circumstances, as is the resulting risk of large-scale fires and the likelihood of cascading effects with severe secondary consequences. This also raises questions about the risk of medium- to long-term soil contamination, in particular if the released substances are toxic or environmentally persistent (Bird and Grossman, 2011). While no detailed study on the consequences of hazardous-substance releases due to tsunami exists, by analogy with river floods it is likely that some substances would react with the tsunami waters and thereby create other chemical compounds that could be even more toxic or flammable.

3.5 FLOODS

River floods can affect individual equipment but also entire hazardous installations through buoyancy and drag forces. Flotation of equipment off its foundations due to flooding of catch basins, as well as subsequent displacement due to water drag, can strain or break tank-pipe connections, leading to minor hazmat leaks but also to more severe, continuous releases (Cozzani et al., 2010). This is a particular problem for empty or nearly empty storage tanks in case anchoring is inadequate or completely missing. While an empty tank by itself does not pose a Natech risk, if it starts floating and is displaced it can become a collision hazard for other equipment onsite. If the force of the floodwaters is sufficiently high, it can cause tanks to collapse or implode, thereby releasing the complete inventory of hazardous substances contained in the unit. Water intrusion in electrical equipment and subsequent power failure or short circuits can affect process and storage conditions and indirectly trigger a Natech accident, for example, via loss of cooling and pressure build up in vessels and emergency flaring of hazardous substances.

Another important hazmat release mode is the impact of floating debris dragged along with the floodwaters on sensitive equipment. This is very similar to debris impact during tsunamis, although the water speed is generally lower during river floods. Depending on the flood height and speed, these floating objects can be smaller items, such as branches, cylinders, or barrels, but also cars or vessels which can do a lot of damage to hazardous facilities. Debris is also a major issue for overland pipelines during flood conditions and an important cause of pipe ruptures. At river crossings, flood-induced erosion and scouring of the river bed can uncover buried pipelines and undermine their foundations, leaving them exposed to debris impact and water pressure (Fig. 3.5). In addition, if too much length of pipe is undermined, it might

FIGURE 3.5 Exposed Petroleum Pipeline Due to Flooding in Pennsylvania, USA

Photo credit: C. Kafer.

break due to the unsupported weight. An in-depth analysis of hazardous liquid pipeline Natech accidents in the United States indicates that the highest amounts of hazmats were released during pipe ruptures caused by floods as opposed to other types of natural hazards (Girgin and Krausmann, 2015).

Like tsunamis, floods usually affect large areas and can carry released hazardous materials over great distances. Consequently, there is the additional risk of the floodwaters becoming a vector for the dispersion of toxic or flammable substances over wide stretches of land with the associated elevated risk of cascading events (Fig. 3.6). Interestingly, floods can also lift flammable waste from an installation's sewer system when the drainage of waste and surface water are not segregated. Contact of these substances with hot surfaces or ignition by a lightning strike can cause fires that spread with the floodwaters (Krausmann et al., 2011; Cruz et al., 2001).

Depending on the hazardous substances involved in the accident, the consequences can be manifold. Nevertheless, water contamination is the most common outcome of flood-triggered Natech events and can include the pollution of surface and underground water, as well as extensive soil contamination in the inundated areas. The severity of the consequences strongly depends on the amount of substances released, their solubility and density (Cozzani et al., 2010). Ignition of flammable substances stratified on the floodwaters, explosions, and the atmospheric dispersion of toxic materials are also common.

FIGURE 3.6 Hydrocarbon Release at a Refinery During the Coffeyville Floods in the USA in 2007

The flood caused a spill of 90,000 gallons of crude oil that polluted a wide swath of land.

Photo credit: Kansas Wing of the Civil Air Patrol.

The postaccident analyses also highlighted additional scenarios ordinarily not considered in conventional industrial accidents. Some chemicals react violently with water and create toxic or flammable vapors in the process. This adds to the risks of the primary substance release by creating a secondary hazard to the population and the environment, but also to emergency responders. Examples of such chemicals are calcium carbide which after contact with water forms flammable acetylene, or cyanide salts which react with water to create hydrogen cyanide, a toxic gas (Cozzani et al., 2010).

3.6 STORMS

Storms comprise a number of phenomena, each of which can adversely affect the chemical-process industry or hazmat-storage areas. These phenomena can cause damage on-shore via flooding from storm surge, that is, high-speed water driven by gale-force winds, and through wind pressure and wind-driven rain (McIntyre and Ford, 2009; Godoy, 2007; Cruz et al., 2001). Offshore facilities are not only affected by high winds but also by wave loading. Blackburn and Bedient (2010) note that storm surge and wave elevations are significantly underrepresented in the engineering literature.

Although no systematic analyses of storm-triggered Natech accident data exist, damage and failure modes can be deduced from single accidents. Flood impacts due to storm surge or heavy rain associated with storms lead to the same types of damage as for river floods discussed in the previous section. Equipment flotation and displacement due to storm surge and wind, the associated breaking of vessel-pipe connections, short circuits, and power outages can cause small but also major chemical releases (Fig. 3.7). Heavy rain can tip or sink floating tank roofs, thereby exposing the tank contents to the air and making them easily accessible to potential ignition sources, such as lightning. Since storm surge can cause a rise in water levels of canals or rivers connected to the ocean, also installations not sited directly at the coast can be subject to storm-surge effects (Cruz and Krausmann, 2013).

Wind-related damage includes shell buckling, toppling of process units and tanks, and tank-roof damage (Godoy, 2007; Cruz et al., 2001). Bailey and Levitan (2008) note that larger tanks have a tendency to fail due to inward collapse of the tank shell, in particular if they are nearly empty, while smaller tanks are more prone to shell buckling after being hit by wind-borne debris, or toppling over. Tank roof damage is caused by the uplift forces of strong winds which exceed the weight of the roof plate (Carson, 2011). For fixed roofs this can result in the tearing of roof-to-shell joints, peeling away of the roof plate, and dislodging of the roof structure (Braune, 2006). Fig. 3.8 gives an example of roof destruction during Hurricane Katrina. In the case of floating roofs, the wind pressure can cause water on the roof to shift, thereby creating an unsymmetrical load which may result in the structural failure of the roof.

High winds, storm surge, and underwater currents can also affect offshore oil and gas infrastructure causing rig tie-down problems and wave loading on platform decks,

FIGURE 3.7 The Water and Gale-force Winds Accompanying Hurricane Katrina Floated and Displaced Two Half-Filled Tanks Along the Mississippi River With 3.8 Million Gallons of Oil Pouring From the Tanks

Photo credit: Louisiana Department of Environmental Quality.

the breaking of platform-riser connections, and mooring failure with subsequent loss of station keeping that can set mobile offshore drilling units adrift (Cruz and Krausmann, 2008; Det Norske Veritas, 2007; Energo Engineering, 2007). Anchor dragging of drifting drilling units and submarine landslides can damage the underwater pipeline network and other subsea facilities.

The type of design and filling level determine the susceptibility of equipment to storm surge or wind impact. Minor, moderate, but also major releases of toxic, flammable, or explosive chemicals are possible both on- and offshore (Cruz and Krausmann, 2009; Cruz et al., 2001). These substances disperse in the air, stratify on or dissolve in water, and are carried away by the flood once the catch basin is overtopped by storm surge or rain-related flooding. Like for floods and tsunamis, the presence of water can trigger chemical reactions that generate additional toxic or flammable substances. During storm conditions the control of hazmat releases from hazardous installations is extremely challenging due to potentially widespread storm damage with power outages, disruption of communication, destruction or blocking of access roads, and the absence of workers displaced by a major storm.

FIGURE 3.8 Wind-Induced Destruction of a Fixed Tank Roof at an Oil Refinery During Hurricane Katrina

Photo credit: NOAA.

3.7 LIGHTNING

Several studies indicate that lightning is one of the most frequent natural-hazard accident triggers in chemical processing and storage activities (Krausmann and Baranzini, 2012; Chang and Lin, 2006; Rasmussen, 1995). In pipeline systems, lightning might be a more important accident initiator than previously thought (Kinsman and Lewis, 2000). It is interesting to note that the effectiveness of commonly implemented lightning protection measures, such as grounding of equipment, lightning rods, or circuit breakers, in preventing fires or damage in industry has proven to be inconclusive (Goethals et al., 2008).

The analysis of lightning-triggered Natech accidents showed that there exist different mechanisms for equipment damage and failure (Renni et al., 2010; ARIA, 2007). Direct structural damage is caused by thermal heating from the lightning strike. This can lead to the puncturing and rupture of tank shells and pipelines. If the lightning energy is not sufficient to pierce the pipe body, it might still disable the pipeline's cathodic corrosion protection system and cause pitting. This spot can become the source of corrosion and failure months after the lightning strike. Indirect structural damage can occur via the collapse of structural components (e.g., flare stacks) struck by lightning that damage hazardous equipment when falling. Another often underestimated indirect damage mechanism is lightning impact on the power grid, or on electrically operated control and safety systems whose disruption could create process upsets and subsequent hazmat releases, for example, from vent and blow-down systems (Renni et al., 2010). This can turn out to be a particular problem during start-up of a hazardous installation following a thunderstorm.

Flammable vapors are often present at the rim seal of floating roof tanks. If lightning strikes the tank roof or in its vicinity, immediate ignition can occur (Fig. 3.9). The energy carried by lightning can also ignite flammable gaseous releases from other types of equipment or stratified substance spills on the ground. Fires are therefore a frequent consequence of lightning strikes involving flammable substances. In fact, fires and explosions are the most frequent outcome of lightning impact for storage tanks. A statistical analysis of selected accident data yields an ignition probability of 0.82 which is not surprising considering that lightning is an ignition source itself. As for earthquakes, this number is likely an upper limit due to the reporting bias toward high-consequence accidents in industrial accident databases (Krausmann et al., 2011).

FIGURE 3.9 Lightning-Triggered Fire in a Gasoline Storage Tank in Oklahoma

Photo credit: Assistant Chief D. Brasuell (ret.), Bixby (OK) Fire Department;
Courtesy of Industrial Fire World Magazine.

The majority of lightning-triggered Natech accidents seem to result in hazmat releases that do not ignite or explode. However, once loss of containment has occurred, the released chemicals can disperse in the air, or cause water and soil contamination. Furthermore, lightning-triggered release quantities can be significant, with almost 40% of the accidents analyzed in Renni et al. (2010) exceeding 1000 kg. Matters are further complicated during lightning storms accompanied by heavy rain which can cause releases from spills in catch basins that overflow or from drainage-water segregation systems whose capacity is exceeded.

3.8 OTHERS

3.8.1 Extreme Temperatures

Recent studies showed that *low temperatures* were among the top three causes of Natech accidents in the European Union and OECD between 1990 and 2009, together with lightning and floods (Krausmann and Baranzini, 2009, 2012). These studies also found that while cold weather and freeze pose an important threat to hazardous installations, the associated risk is in most cases severely underestimated.

During conditions of extreme or prolonged cold weather with temperatures below zero, or quickly alternating freeze and thaw phases, various types of industrial equipment and their appurtenances are vulnerable to malfunction and damage. This includes pipework (e.g., transfer or drainage lines) that is not adequately insulated or lacks insulation, pumps, valves, and flanges on tanks or pipes, pipe welds and joints, but also control systems and sensors performing safety-relevant functions (ARIA, 2012a; CSB, 2008).

There are different mechanisms by which low temperatures can cause damage in hazardous installations. The freezing of industrial equipment or pipeline network components can lead to component malfunction and leaks, for instance due to valve or control-system failures (e.g., overfilling caused by a frozen level detector). Ice formation in pipes can cause blockage of auxiliary pipes and overpressure which might eventually lead to pipe rupture, or tank overflow if an overflow line is blocked by ice. In the pipeline network the most important cold-weather related damage mode is the expansion of freezing water and subsequent pipe cracking or bursting by mechanical forces. Water naturally present in the transported hazardous substances is the main source of the ice. Another important damage mechanism is frost heave in which a section of the ground is lifted upward due to the freezing of water present in the soil. If a buried pipeline crosses this area, it will be affected by the vertical ground displacement and can buckle. Falling ice and snow can create physical loads on the equipment that could also cause cracks and hazardous-materials releases (Girgin and Krausmann, 2015).

Although severely underestimated as an accident initiator, cold-weather related damage to hazardous equipment is frequent and has caused ground, water, and atmospheric pollution, as well as fires and explosions with sometimes significant emergency-response, cleanup, and restoration costs.

Hot weather can affect hazardous installations in different ways. A recent analysis of heat-related accidents in France concluded that direct exposure to solar radiation and thermal stresses can cause reactive substances to evaporate and self-ignite (ARIA, 2012b, 2015). High temperatures can lead to the decomposition and polymerization of substances, or the formation and accumulation of flammable vapors in confined spaces with an elevated risk of ignition. In pressurized systems, excessive heat can trigger pressure surges with the subsequent actuation of safety valves and leaks of hazardous gases. Auxiliary equipment, such as pumps or compressors, can also be a source of releases and spills if they overheat and break down. In pipeline systems, Girgin and Krausmann (2015) identified heat-induced valve and strainer failures, as well as collar joint failures due to ground shifts brought about by extended drought periods. While pollution and explosions have been observed in relation to hot weather conditions, fires appear to be the predominant outcome.

3.8.2 Volcanoes

There are very few studies related to Natech risks due to volcanic eruptions. Milazzo et al. (2013a,b) and Salzano and Basco (2009) analyzed the danger to industrial facilities around Mt. Etna and Mt. Vesuvius volcanoes in Italy. These areas are characterized by strong seismic activity triggered by the volcanoes, a high population density, and intense industrialization, including large industrial ports.

These studies focused on the potential risks to atmospheric storage tanks due to volcanic ash fallout which is considered to be an important phenomenon potentially affecting large areas by which the integrity of industrial facilities could be compromised. In this context, the main damage and failure mechanisms are an increase in the loading on fixed or floating tank roofs due to ash accumulation, or damage to rubber seals caused by the abrasiveness of the ash. Assessing the resistance of fixed roofs to damage by ash loading, and of floating roofs to sinking or capsizing, Milazzo et al. (2013a) estimated threshold values for ash deposits for both wet and dry conditions. Saturation of volcanic ash with rainwater and the associated increase in density reduce the ash load required to cause damage to atmospheric tank roofs. Overall, the study showed that only extremely large explosive eruptions could damage tank roofs or cause them to sink or capsize. Simple mitigation actions, such as removing the ash by blowing or brushing it from the roof, can be effective.

Another volcano-related phenomenon that can endanger industrial facilities and lead to hazardous-materials releases are lahar flows. Lahars are volcanic mudflows triggered by the melting of ice and snow that while descending the volcano can damage or bury infrastructures in exposed areas. During the 1989/1990 eruptions of Redoubt Volcano in Alaska, lahar flows caused partial flooding of a crude-oil terminal in the Drift River valley. Although no oil was spilled, in the aftermath of the event the operator of the terminal built dikes of 6-m height around almost all the terminal and raised critical electrical equipment to a minimum of 1-m above the ground (Brantley, 1990). The oil terminal was again affected by lahars during Redoubt's 2009 eruption, but the dikes around the tank farm effectively protected it from a potentially catastrophic inundation and oil spills (Fig. 3.10). Pierson et al. (2014) argue that exclusion dikes can effectively enclose and protect valuable infrastructure from lahar impact.

FIGURE 3.10 View West of the Oil Terminal in Drift River Valley, Alaska

Lahar deposits have ramped up to the top of the west dike and spilled over in a couple of locations. Water entered the compound along an existing roadway in the foreground.

Photo credit: G. McGimsey, AVO/USGS.

References

Antonioni, G., Bonvicini, S., Spadoni, G., Cozzani, V., 2009. Development of a framework for the risk assessment of Natech accidental events. Reliab. Eng. Syst. Saf. 94, 1442.

Arcidiacono, V., Krausmann, E., Girgin, S., 2014. Seismic vulnerability of chemical racks in the cross-aisle direction, EUR 26953 EN, European Union.

ARIA, 2007. Lightning—Elements of industrial accidentology, France 1967–2007, Bureau for analysis of industrial risks and pollutions (BARPI). Ministry of Ecology, Sustainable Development and Energy, France. http://www.aria.developpement-durable.gouv.fr/analyses-and-feedback/by-theme/lightning-industrial-accidentology/?lang=en

ARIA, 2012a. Intense cold weather: beware of freezing… and then thawing!, ARIA Newsflash November 2012. Ministry of Ecology, Sustainable Development and Energy, France.

ARIA, 2012b. Heat waves, intense heat: risk alerts, and not just for fire!, ARIA Newsflash May 2012. Ministry of Ecology, Sustainable Development and Energy, France.

ARIA, 2015. Results of an accident study covering the 2015 summer heat wave in France, December 2015. Ministry of Ecology, Sustainable Development and Energy, France.

Bailey, J.R., Levitan, M.L., 2008. Lessons learned and mitigation options for hurricanes. Process Saf. Prog. 27 (1), 41.

Bird, W.A., Grossman, E., 2011. Chemical aftermath—contamination and cleanup following the Tohoku earthquake and tsunami. Environ. Health Perspect. 119 (7), A290.

Blackburn, J., Bedient, P., 2010. Learning the lessons of Hurricane Ike, Severe Strom Prediction, Education and Evacuation from Disasters (SSPEED) Center. Rice University, Texas.

Brantley, S.R. (Ed.), 1990. The eruption of Redoubt Volcano, Alaska, December 14, 1989—August 31, 1990. U.S. Geological Survey Circular 1061.

Braune, S.L., 2006. Preparing tanks for hurricanes, API 2006 Storage Tank Management and Technology Conference. Tulsa, Oklahoma, USA, September 25–28.

Bureau, G., Kokkas, P., 1992. Seismic retrofit of oil storage tanks. In: Proceedings of the Tenth World Conference on Earthquake Engineering. Madrid, Spain, July 19–24.

Campedel, M., 2008. Analysis of major industrial accidents triggered by natural events reported in the principal available chemical accident databases, EUR 23391 EN, European Communities.

Carson, G., 2011. Encoding safety, Hydrocarbon Engineering. Palladian Publications, UK, June.

Chang, J.I., Lin, C.C., 2006. A study of storage tank accidents. J. Loss Prev. Process Ind. 19 (1), 51.

Cozzani, V., Campedel, M., Renni, E., Krausmann, E., 2010. Industrial accidents triggered by flood events: analysis of past accidents. J. Hazard. Mater. 175, 501.

Cruz, A.M., Krausmann, E., 2008. Damage to offshore oil and gas facilities following Hurricanes Katrina and Rita: an overview. J. Loss Prev. Process Ind. 21 (6), 620.

Cruz, A.M., Krausmann, E., 2009. Hazardous-materials releases from offshore oil and gas facilities and emergency response following Hurricanes Katrina and Rita. J. Loss Prev. Process Ind. 22 (1), 59.

Cruz, A.M., Krausmann, E., 2013. Vulnerability of the oil and gas sector to climate change and extreme weather events. Climatic Change 121, 41.

Cruz, A.M., Steinberg, L.J., 2005. Industry preparedness for earthquakes and earthquake-triggered hazmat accidents during the Kocaeli earthquake in 1999: a survey. Earthquake Spectra 21, 285.

Cruz, A.M., Krausmann, E., Selvaduray, G., 2016. Hazardous materials: earthquake-triggered incidents and risk reduction approaches. In: Scawthorn, C., Chen, W.-F. (Eds.), Earthquake Engineering Handbook. second ed. CRC Press LLC, Boca Raton.

Cruz, A.M., Steinberg, L.J., Luna, R., 2001. Identifying hurricane-induced hazardous materials release scenarios in a petroleum refinery. Nat. Hazards Rev. 2 (4), 203.

CSB, 2008. LPG fire at Valero-McKee Refinery. U.S. Chemical Safety and Hazard Investigation Board, Investigation Report No. 2007-05-I-TX.

Det Norske Veritas, 2007. Pipeline damage assessment from Hurricanes Katrina and Rita in the Gulf of Mexico, Report 448 14183, Rev.1.

Energo Engineering, 2007. Assessment of fixed offshore platform performance in Hurricanes Katrina and Rita. Final Report, MMS Project No. 578, Energo Engineering Project No. E06117.

FEMA, 2012. Guidelines for Design of Structures for Vertical Evacuation From Tsunamis—Report P-646. Federal Emergency Management Agency, Redwood City, CA.

Girgin, S., Krausmann, E., 2015. Lessons learned from oil pipeline natech accidents and recommendations for natech scenario development. JRC Science and Policy Report EUR 26913 EN, European Union.

Godoy, L.A., 2007. Performance of storage tanks in oil facilities damaged by Hurricanes Katrina and Rita. J. Perform. Constr. Facil. 21 (6), 441.

Goethals, M., Borgonjon, I., Wood, M. (Eds.), 2008. Necessary measures for preventing major accidents at petroleum storage depots—key points and conclusions. Seveso Inspection Series, vol. 1, EUR 22804 EN, European Communities.

Kinsman, P., Lewis, J., 2000. Report on a Study of International Pipeline Accidents, Contract Research Report 294/2000. HSE Books, Norwich, United Kingdom.

Krausmann, E., Baranzini, D., 2009. Natech risk reduction in OECD Member Countries: results of a questionnaire survey, Report JRC 54120, European Communities, limited distribution.

Krausmann, E., Baranzini, D., 2012. Natech risk reduction in the European Union. J. Risk Res. 15 (8), 1027.

Krausmann, E., Cruz, A.M., 2013. Impact of the 11 March 2011, Great East Japan earthquake and tsunami on the chemical industry. Nat. Hazards 67, 811.

Krausmann, E., Cruz, A.M., Affeltranger, B., 2010. The impact of the 12 May 2008 Wenchuan earthquake on industrial facilities. J. Loss Prev. Process Ind. 23, 242.

Krausmann, E., Renni, E., Cozzani, V., Campedel, M., 2011. Major industrial accidents triggered by earthquakes, floods and lightning: results of a database analysis. Nat. Hazards 59 (1), 285–300.

Lanzano, G., Salzano, E., Santucci De Magistris, F., Fabbrocino, G., 2014. Seismic vulnerability of gas and liquid buried pipelines. J. Loss Prev. Process Ind. 28, 72.

Lanzano, G., Santucci De Magistris, F., Fabbrocino, G., Salzano, E., 2015. Seismic damage to pipelines in the framework of Na-Tech risk assessment. J. Loss Prev. Process Ind. 33, 159.

Lopez, O.A., Grases, J., Boscan, L., Rojas, M., 1992. Seismic retrofitting of oil refinery equipment. Madrid, Spain. Proceedings of Tenth World Conference on Earthquake Engineering., July 19–24.

MAHB, 2014. Lessons Learned Bulletin No. 6—Natech accidents, Major Accident Hazards Bureau, European Commission, JRC 93386.

McIntyre, D.R., Ford, E., 2009. Recent developments in the analysis of fires, explosions, and production disruption incidents in chemical plants and oil refineries. Process Saf. Prog. 28 (3), 250.

Milazzo, M.F., Ancione, G., Basco, A., Lister, D.G., Salzano, E., Maschio, G., 2013a. Potential loading damage to industrial storage tanks due to volcanic ash fallout. Nat. Hazards 66, 939.

Milazzo, M.F., Ancione, G., Salzano, E., Maschio, G., 2013b. Risks associated with volcanic ash fallout from Mt. Etna with reference to industrial filtration systems. Reliab. Eng. Syst. Saf. 117, 73.

Nishi, H., 2012. Damage on hazardous materials facilities. Tokyo, Japan. Proceedings of the International Symposium on Engineering Lessons Learned from the 2011 Great East Japan Earthquake., March 1–4, 2012.

Pawirokromo, J., 2014. Design of a flood proof storage tank, Stolt-Nielsen terminal, New Orleans. Masters Thesis, Delft University of Technology.

Pierson, T.C., Wood, N.J., Driedger, C.L., 2014. Reducing risk from lahar hazards: concepts, case studies, and roles for scientists. J. Appl. Volcanol. 3, 16.

Rasmussen, K., 1995. Natural events and accidents with hazardous materials. J. Hazard. Mater. 40 (1), 43.

Renni, E., Krausmann, E., Cozzani, V., 2010. Industrial accidents triggered by lightning. J. Hazard. Mater. 184, 42.

Salzano E., Basco A., 2009. A preliminary analysis of volcanic Na-Tech risks in the Vesuvius area, In: Martorell, S., Guedes Soares, C., Barnett, J. (Eds), Safety, reliability and risk analysis: theory, methods and applications. In: Proceedings of the ESREL Conference, Valencia, Spain, September 22–25, 2008. Taylor & Francis Group, London.

Salzano, E., Basco, A., 2015. Simplified model for the evaluation of the effects of explosions on industrial target. J. Loss Prev. Process Ind. 37, 119.

Salzano, E., Garcia Agreda, A., Di Carluccio, A., Fabbrocino, G., 2009. Risk assessment and early warning systems for industrial facilities in seismic zones. Reliab. Eng. Syst. Saf. 94, 1577.

Zama, S., Nishi, H., Yamada, M., Hatayama, K., 2008. Damage of oil storage tanks caused by liquid sloshing in the 2003 Tokachi Oki earthquake and revision of design spectra in the long-period range. Beijing, China. Proceedings of the Fourteenth World Conference on Earthquake Engineering., October 12–17.

Zama, S., Nishi, H., Hatayama, K., Yamada, M., Yoshihara, H., Ogawa, Y., 2012. On damage of oil storage tanks due to the 2011 off the Pacific Coast of Tohoku Earthquake (Mw9.0), Japan. In: Proceedings of the Fifteenth World Conference on Earthquake Engineering, Lisbon, Portugal, September 24–28.

Status of Natech Risk Management

E. Krausmann*, R. Fendler**, S. Averous-Monnery†, A.M. Cruz‡, N. Kato§

*European Commission, Joint Research Centre, Ispra, Italy; **German Federal Environment Agency, Dessau-Roßlau, Germany; †United Nations Environment Programme (UNEP), Division of Technology, Industry and Economics, Paris Cedex, France; ‡Disaster Prevention Research Institute, Kyoto University, Kyoto, Japan; §Department of Naval Architecture and Ocean Engineering, Osaka University, Osaka, Japan

In many countries, a legal framework for chemical-accident prevention and mitigation exists and some programs address Natech risk directly or implicitly. Nonetheless, the repeated occurrence of Natech accidents raises questions about the effectiveness of these frameworks. This chapter provides examples of current national approaches and international activities to manage Natech risks.

4.1 REGULATORY FRAMEWORKS

4.1.1 European Union

In the European Union (EU), major (chemical) accident risks are regulated by the provisions of the so-called Seveso Directive on the control of major-accident hazards (European Union, 1997) and its amendments (European Union, 2003, 2012). The Seveso Directive applies to industrial activities that use, handle, or store specific types of hazardous substances beyond a certain threshold quantity, and certain hazardous substances considered to be extremely dangerous. The Directive does not cover all types of risks from activities that involve hazardous materials. Risks from pipelines, landfills, military or nuclear installations, or the transport of hazardous goods are excluded.

The Seveso Directive requires stringent safety measures to be put in place to prevent major accidents from occurring and in case they do happen to effectively limit their consequences for human health and the environment. For instance, the Directive requires the drawing up of a major-accident prevention policy (MAPP) by the operator, and for particularly hazardous plants (so-called upper-tier establishments) a safety report to ensure high protection levels. Through the safety report, the operator is obliged to demonstrate that (1) a MAPP has been implemented, (2) major-accident hazards and associated accident scenarios have been identified and necessary measures have been taken to prevent these accidents and limit their consequences if they nevertheless occur, (3) adequate safety and

Natech Risk Assessment and Management. http://dx.doi.org/10.1016/B978-0-12-803807-9.00004-8

reliability have been taken into account in the design, construction, operation, and maintenance of the installations, and (4) internal emergency plans have been drawn up and provide sufficient information for the preparation of the external emergency plan. The Seveso Directive also addresses domino effects by requiring the identification of establishments for which the risk of a major accident may be increased due to the geographical position and the proximity of these establishments, and their inventories of hazardous substances. Provisions on land-use planning control the siting of new or the modification of existing hazardous establishments, or new developments in their vicinity.

Following a series of Natech and other major chemical accidents (e.g., the spill of cyanide-laced tailings from a dam breach in Romania due to heavy rainfall and rapid snowmelt, or the release of chlorine from a flooded general-chemicals manufacturer), it was decided that an amendment of the Seveso Directive was needed to address the remaining gaps. The latest amendment, which entered into force in 2012, now explicitly addresses Natech risks and requires that environmental hazards, such as floods and earthquakes, be routinely identified and evaluated in an industrial establishment's safety report (European Union, 2012). Awareness of Natech risks has been growing ever since in Europe.

From the point of view of Natech risk reduction, the Seveso Directive is the most important legal act at EU level. Other regulations exist that control different industrial activities or areas of concern that sometimes indirectly address Natech risks. One example is the EU Water Framework Directive, which aims to establish an integrated river basin management in Europe and which includes provisions against the chemical pollution of surface water (European Union, 2000). The EU Floods Directive requires Member States to develop flood risk maps that shall also show hazardous installations, which might cause accidental pollution in case of flooding (European Union, 2007). With respect to offshore oil and gas infrastructure, the recent EU Offshore Directive requires a demonstration that all major hazards related to offshore activities are identified, their likelihood and consequences assessed, and appropriate risk-control measures put in place. This includes environmental and meteorological conditions that could pose limitations on safe operations (European Union, 2013).

4.1.2 United States of America

There are several federal programs in place for hazardous-materials risk management and emergency-response planning in the United States including the Process Safety Management (PSM) regulation and Risk Management Plan (RMP) rule (OSHA, 2012; EPA, 2015). In response to these requirements, industrial facilities, in addition to carrying out a process-safety analysis, need to maintain process-safety information, evaluate existing mitigation measures and standard operating procedures, and develop training and maintenance programs. An emergency-response program is also required which should clearly describe the measures taken to protect human health and the environment in response to an accidental hazardous-materials release and the procedures for notifying the public and local agencies in case of a chemical emergency. However, neither of these federal regulations explicitly require analyzing, preparing for, or mitigating releases which are concurrent with natural disasters. Furthermore, there are no specific provisions in the PSM or the RMP to prevent domino effects, which occur more frequently during earthquakes, or for land-use planning. However, some states have adopted stricter legislation. For instance, due to the high risk of earthquakes in California, the California Accidental Release Prevention (CalARP) Program calls specifically for a risk assessment of potential releases due to an earthquake and the adoption of appropriate prevention and mitigation measures to prevent the release of certain hazardous substances during earthquakes (CalARP, 2014; Cruz and Okada, 2008).

4.1.3 Japan

In Japan, the control of major chemical-accident risks is regulated by various laws, including the High Pressure Gas Safety Law, the Fire Service Law, and the Law on the Prevention of Disasters in Petroleum Industrial Complexes and Other Petroleum Facilities (also Petroleum Complex Disaster Prevention Law). None of these laws specifically requires carrying out a risk assessment for potential chemical accidents with off-site consequences (High Pressure Gas Safety Institute of Japan, personal communication). However, damage incurred by petroleum refineries during past earthquakes has prompted the adoption of a broad range of earthquake risk-reduction measures. The Petroleum Complex Disaster Prevention Law was updated in the wake of the 2003 Tokachi-oki earthquake, which triggered a large-scale tank fire. It now stipulates the implementation of earthquake-specific safety measures for floating roof tanks and targeted fire-fighting strategies in case of major tank fires (PAJ, 2011; CAO, 2012). The amended Fire Service Law regulates the storage, handling, and use of nonpressurized toxic materials and flammables (CAO, 2008) while the amended High Pressure Gas Safety Law controls high-pressure gases and liquefied pressurized gases (HPGSIJ, 2005). Cruz and Okada (2008) note that the High Pressure Gas Safety Law is the only regulation that explicitly addresses Natech risks in industrial establishments by requiring that measures be taken to reduce the risk of accidents from earthquakes and tsunamis. Companies have to prepare for the eventuality of an accident in compliance with the legislation in force that covers their industrial activity.

Since the Great East Japan Earthquake in 2011, the seismic code for high-pressure gas storage facilities has been improved to account for the effects of long-period seismic motion-induced liquid sloshing on storage tanks. Furthermore, the new code increases the seismic resistance capacity of the supporting frames of pipe braces by reinforcing the intersection of the braces.

In addition, a new Land Resilience Basic Law was introduced in 2013 with the goal to promote long-term societal resilience through sustainable construction design, as well as ensuring land resilience through improved ground conditions and design. More specifically, the new Law requires the adoption of comprehensive countermeasures against (1) destruction, fire, and explosion at industrial complexes, (2) operational failures or disruption of the supply chain of oil and LPG, and (3) occurrence of an extended and complex disaster in highly populated bay areas (e.g., Tokyo Bay, Osaka Bay) caused by industrial-installation damage. The countermeasures against (1) include an amendment of the regulation concerning underground storage tanks, and the publication of a seismic risk-assessment guideline for industrial facilities with hazardous materials. Those against (2) include a guideline on storage and transport of hazardous materials to ensure smooth operations in case of an emergency, and an amendment of regulations concerning the shipping and storage of hazardous materials by ship and tanker truck also to guarantee well-ordered operations during emergency conditions.

4.1.4 Colombia

In response to the growing losses from natural and technological disasters, the Colombian government passed a new Law, entitled "Policy for Disaster Risk Reduction" in Apr. 2012 (Law 1523, April 2012). The new law requires the management of all kinds of risks, including natural, man-made, and Natech risks, in order to implement risk-reduction actions. The new law transforms the old system into a new National System for Disaster Risk Management (SNGRD). Following the passing of the new law, the National Unit for Risk Management of Disasters (UNGRD) was established. This new law is important because it specifically considers both natural and technological hazards as sources of potential disasters, and requires coordinated actions by all stakeholders, in particular concerning land-use planning.

In the frame of the new law, an alliance between the United Nations Development Program (UNDP) and the UNGRD was established to respond to the need to strengthen capacities at the local and national levels, and to improve territorial planning. Furthermore, a public–private partnership was established between UNDP and the Colombian Petroleum Company S. A. (Ecopetrol S. A.). The aim of the partnership was to contribute to the design and implementation of a national strategy for the promotion of the safe and sustainable coexistence between the transport of hydrocarbons (there are over 8000 km of oil and high-pressure gas pipelines in Colombia), the territory, and the inhabitants living in or near the pipeline corridors. Thus, one of the main objectives of the project was the incorporation of technological risk-management instruments in local and national territorial planning (Cruz, 2014).

4.2 IMPLEMENTATION OF NATECH RISK REDUCTION

4.2.1 European Union

EU Member States have to transpose the provisions of the Seveso Directive into national law. The Directive does not prescribe how these goals should be reached. It thereby gives some freedom to the Member States to choose the approach they consider most appropriate for achieving the objectives stipulated in the Directive. As an example, Section 4.2.2 discusses in detail how the Seveso Directive has been transposed into German law.

A recent survey among Seveso regulatory bodies aimed to assess the status of Natech risk management in the EU, collect case histories and lessons learned, and identify needs and/or limitations in implementing Natech risk-reduction strategies in EU Member States (Krausmann and Baranzini, 2012; Krausmann, 2010). The results of the survey were encouraging as they showed an increasing awareness among the competent authorities of the potentially disastrous impacts of natural hazards on chemical facilities. However, the survey also highlighted a number of gaps in Natech risk reduction, as well as related research and policy challenges.

Over half of the survey respondents indicated that their countries had experienced Natech accidents, which sometimes resulted in fatalities and injuries in the period 1990–2009. The main natural hazards of concern in this context were lightning, low temperatures, and floods. Considering the recurrence of Natech accidents, the survey results suggest that the legal frameworks for chemical-accident prevention have not always been effective. The reasons may be manifold and include, for instance, the absence of systematic data collection and analysis, which has resulted in incomplete knowledge of the dynamics of Natech accidents. This is turn has led to a lack of equipment vulnerability models for most natural hazards, and of methodologies and scenarios for considering Natech risks in industrial risk assessment. As a consequence, the survey participants expressed their belief that industry in many EU countries may not consider Natech risks appropriately in their facility risk assessment, with potentially low preparedness levels as a result. The survey also revealed strong differences between the actual Natech accident triggers and the natural hazards perceived to be of concern in the participating countries, highlighting an incongruity between actual causes and risk perception.

The recurrence of Natech accidents has also cast doubt on the adequacy of the design basis of hazardous installations with respect to natural-hazard impact, as well as on the associated protection measures in place. Design codes and standards usually assume a cut-off limit for the intensity of a reference natural hazard against which an installation and its components are designed. However, it is unclear which level of damage or even failure is to be expected above the design-basis loading. Moreover, the survey respondents emphasized that the ultimate objective of these codes and standards is the preservation of life safety and hence the prevention of building collapse. While in itself an important goal, the preservation of a building's structural integrity is not necessarily sufficient to prevent the release of hazardous materials under natural-event loading. Failure to recognize these design limits and the specific requirements of process equipment during natural-hazard impact could have contributed to diminishing the effectiveness of chemical-accident prevention frameworks in reducing Natech risks.

The survey identified a number of key areas for future work for industry, regulators, and science and engineering. The majority of survey respondents called for the development of guidance on Natech risk assessment for industry with the highest priority, followed by the preparation of Natech risk maps to inform land-use- and emergency-planning by identifying a region's Natech hot spots. Work is underway toward closing these gaps, although data availability remains a challenge.

The survey on the status of Natech risk management in the EU focused on polling Seveso competent authorities, recognizing the major accident potential of Seveso-type installations. However, also industrial installations processing, handling or storing hazardous materials in quantities below the qualifying threshold of the Seveso Directive can pose a significant safety risk. These industrial activities are governed by legal acts whose safety provisions may be less strict than those stipulated by the Seveso Directive. This is an additional factor that may have contributed to hampering effective Natech risk reduction.

4.2.2 Germany

The control of major accident risk was already subject to the German Federal Immission Control Act (BImSchG) first issued in 1974 (BRD, 2015). The major accidents at Feyzin in Jan. 1966, Flixborough in Jun. 1974, Seveso in Jul. 1976, and Manfredonia in Sep. 1976 triggered a discussion on the need for a more detailed regulation to control the risks of installations where large amounts of hazardous substances are or may be present. One result was the German Major Accident Ordinance based on the BImSchG that entered into force in Sep. 1980. Since 1980 this ordinance has been amended several times and implements now the main part of the EU Seveso Directive. A revision for the implementation of the requirements of the 2012 amendment of the Seveso Directive is intended for 2016.

The German Major Accident Ordinance currently includes six main types of obligations of operators:

1. basic general requirements,
2. technical and organizational safety requirements,
3. elaboration and implementation of a safety management system,
4. elaboration of a safety report and emergency plan (both "upper tier" obligations),
5. information on risks, safety, and behavior in case of emergency (an "upper tier" obligation),
6. notification of accidents.

The basic general requirements in Section 3 of the Major Accident Ordinance include the obligations of operators:

a. To take such precautions according to the nature and extent of the potential hazards as are necessary to prevent major accidents.
b. To take precautions to keep the effects of (nevertheless occurring) major accidents as small as possible.
c. The nature and operation of the installations in the establishment must be according to the state-of-the-art of safety technology.

For the determination of the safety measures required for accident prevention or the limitation of the consequences of (nevertheless occurring) major accidents, the operator has to take into consideration:

a. operational hazard sources,
b. environmental hazard sources, such as earthquakes or floods, and
c. interference by unauthorized persons.

"Environmental hazard sources" include:

1. Technical hazards arising outside the establishment and causing a hazard to the installations in the establishment.
2. Natural hazards arising in the establishment or outside but causing a hazard to the installations in the establishment.

The first point includes, for example, other hazardous installations, such as pipelines for the transport of hazardous materials, dams, and high-voltage power lines, while the second point comprises all natural hazards able to cause a major accident via impacts on the hazardous installations in the establishment.

Due to these obligations in Section 3 of the Major Accident Ordinance, operators and authorities in charge of installations or establishments being subject to the Major Accidents Ordinance have had to take Natech risks into consideration already since 1980. Especially in the evaluation of risks and the determination of the required prevention and preparedness measures in safety reports, natural hazards had to be regarded like operational technical hazards. The consequences of this regulation are five important aspects:

1. The operator is in charge of controlling the Natech risks due to his establishment.
2. The operator has to evaluate which natural hazards may affect his establishment.
3. Natural-hazard maps play an important role for doing this.
4. The operator has to consider natural hazards in his hazard or risk analysis.
5. The operator has to consider natural hazards in his safety or risk management.

In the future, these basic obligations of the Major Accident Ordinance shall remain unchanged and the requirements for safety-management systems and safety reports shall be as stipulated in the amended Seveso Directive, including all requirements for Natech risk management there.

In 2002, the Czech Republic, Austria, and the eastern and southern parts of Germany were affected by a major flood. This flood caused no major accidents in Germany but several minor releases of hazardous substances, mainly heating oil spilled from domestic tanks. In addition, there were at least two "near misses" at establishments according to the Major Accidents Ordinance. In the Czech Republic, the flood caused several spills and a major release of chlorine from a chemical plant. Due to these events the German Environment Agency decided to commission a study on Natech risk management. This study included a review of regulations and technical rules considering Natech, an evaluation of events during the flood in 2002 and of Natech risks at other establishments, and information on the state-of-the-art in Natech risk reduction and recommendations for future activities (Warm and Köppke, 2007). One of the results of this study was that the technical rules applied for Natech risk management in Germany were not developed and hence suitable for establishments. They did not consider the additional risks due to large amounts of hazardous substances present in establishments and therefore the Natech risks from establishments. The authors recommended either to amend the existing rules or to develop special technical rules for Natech risk management with risk-proportional requirements, that is, defining required safety measures according to the Natech risks of establishments.

Another important source of Natech risk-management activities in Germany were discussions on the need to adapt to the effects of climate change. The fourth assessment report of the Intergovernmental Panel for Climate Change (IPCC, 2007) and several studies with regional scope raised awareness not only of the expected effects of climate change but also of the need to adjust to climate change, including adaptation to more frequent or more intense hydrometeorological extreme events. Therefore, the German Government decided to develop a strategy for adaptation to climate change (Bundesregierung, 2008). This strategy also addresses Natech aspects and determines that concerning process safety there shall be adapted:

- regulations and technical rules,
- safety management,
- emergency planning,
- safety measures of establishments due to extreme precipitation and flood events,
- safety measures of establishments due to more frequent or intense storms.

According to this decision of the German Government, the Commission for Process Safety, tasked with the determination of the scientific and technical basis for major-accident prevention and preparedness, prepared two Technical Rules on Installation Safety (TRAS) for Natech Risk Management:

a. TRAS 310: Precautions and Measures against the Hazard Sources Precipitation and Flooding
b. TRAS 320: Precautions and Measures against the Hazard Sources Wind, Snow-, and Iceloads

Both TRAS were put into effect by the Federal Ministry for the Environment, Nature Conservation, Building and Nuclear Safety. The preparation of the TRAS was supported by research projects which provided additional guidance and scientific background information, especially with regard to the consequences of climate change, and which include results of tests made by applying the draft TRAS to installations in establishments (Köppke et al., 2013; Krätzig et al., 2016). These technical rules will be discussed in more detail in Chapter 8.

Further German activities at the national level will be to support the enforcement of both TRAS, and at the international level the raising of awareness of Natech risks and implementation of Natech risk management.

4.3 INTERNATIONAL ACTIVITIES

4.3.1 OECD Guiding Principles for Chemical Accident Prevention, Preparedness and Response

4.3.1.1 The OECD Natech Project

One of the main international guidelines considering Natech risks are the OECD Guiding Principles for Chemical Accident Prevention, Preparedness and Response (OECD, 2003). Elaborated by the OECD Working Group on Chemical Accidents (WGCA) and first published in 1992, the application of the Guiding Principles is the subject of an OECD Council Recommendation. Since the latest revision of the Guiding Principles considered only some aspects of Natech risk management, the WGCA decided to address the issue more comprehensively by including a Natech project into its 2009–12 work program to identify existing gaps and develop targeted recommendations for Natech risk reduction.

As a first step, a questionnaire survey on the status of Natech risk management in OECD Member Countries was carried out in parallel to the survey in EU Member States discussed in Section 4.2.1. The OECD results showed a similar trend as for the EU survey, and highlighted the same gaps (Krausmann and Baranzini, 2009). While there is increasing awareness that natural hazards can be an important external hazard source in hazardous installations, Natech accidents keep occurring. Although comprising different legislative regimes for chemical-accident prevention and preparedness, the majority of OECD survey respondents expressed their belief that there is a clear need for improving current regulations and filling existing gaps to fully address Natech risk reduction. Similar to the EU survey, they called for the development of natural-hazard and Natech risk maps, methodologies for and guidance on Natech risk assessment for industry and communities, as well as training of authorities on Natech risk reduction.

With the OECD Natech survey and a discussion document including recommendations for good practices in Natech risk management as a basis, an OECD Workshop on Natech Risk Management was organized in 2012, with participation from authorities, operators, academia, and NGOs from 19 countries and 5 supranational or international organizations (OECD, 2013). The workshop discussed specific elements of the prevention of, preparedness for, and response to Natech accidents, and a consensus was sought and reached on priority recommendations for Natech risk management. This filtered into the preparation of an addendum to the OECD Guiding Principles to address Natech risk reduction in a systematic and comprehensive manner.

4.3.1.2 The Natech Addendum to the OECD Guiding Principles

The OECD Natech Addendum was published in 2015 and supplements the Guiding Principles with (OECD, 2015):

1. Additions to or modifications of recommendations already included in the OECD Guiding Principles.
2. A separate new chapter with specific recommendations for Natech risk management that elaborates in detail on issues already addressed in the first part.

Both parts of the Natech Addendum include numerous recommendations for government and industry. The recommendations address the inclusion of Natech risks in the drafting of regulations, rules and standards, their enforcement and implementation, and other activities in support of effective Natech risk management. The recommendations are relevant for authorities in charge of the management of risks of hazardous installations, as well as for those dealing with natural-hazard risks. However, due to the approach of the OECD Natech project, the integration of Natech risk management in natural-disaster management is not addressed. With pipelines being at risk due to natural hazards, the Natech Addendum advocates the consideration of Natech risks in pipeline safety, as well.

Like the OECD Guiding Principles, the addendum starts with recommendations for industry. Operators of hazardous installations should take risks due to natural hazards into consideration in the preparation of *safety reports*. The full spectrum of natural hazards able to affect an installation has to be taken into account. In addition, operators should consider naturals hazards in their hazard identification and *risk assessment* for relevant installations. In both cases, operators should recognize that there may be data limitations concerning natural hazards. New information on natural hazards can therefore be one of the reasons for updating safety reports. The Addendum does not specify whether it is the responsibility of the operator or of the authorities to fill relevant data gaps on natural hazards threatening an installation; it recommends that authorities should close the data gaps but this does not change the responsibility of the operator to consider all relevant natural hazards in his risk assessment and safety report.

Many natural hazards and their intensities are linked to a specific area or location. Consequently, operators should take natural-hazard risks into consideration in the *siting* of a new installation or of a significant modification of an existing installation. In this context, operators should consider that climate change may affect the frequency, intensity, and geography of some natural hazards. Regional climate-change projections should therefore be factored in and the Natech risk-management process should be linked to an installation's strategy for climate-change adaptation. A dialog between authorities in charge of chemical-accident risks and those responsible for managing natural risks would be beneficial.

Industry should also consider natural hazards in the *design, planning, and layout* of installations, as well as in *operating procedures*. The standards, codes of practice, and guidance used in design and layout should take into account information on and risks associated with natural hazards. The design and construction of existing installations should be reviewed on the basis of actual knowledge on possible impacts by natural hazards and retrofits made, if required. Moreover, like for other abnormal conditions, the operators should develop special operating procedures for natural-hazard situations. Although not explicitly mentioned by the Natech Addendum, this may include procedures for emergency shutdown during or after extreme natural hazards, and in particular in case of beyond-design basis natural events.

Emergency plans (of operators and public authorities) should take into account natural hazards and the potential occurrence of Natech accidents. Natech risks should be identified in the preparation of scenarios for emergency planning, as well as the zones where effects are likely to occur. Emergency plans should also consider the possible impacts of natural hazards on infrastructure and response capabilities. Warning systems for natural hazards should be developed to provide alerts of an imminent threat to hazardous installations and communities.

The Natech Addendum further recommends that public authorities make arrangements for the development, dissemination, use, and updating of *natural-hazard maps* including all relevant natural hazards in an area. Operators and authorities should also consider natural hazards and Natech risks in their *land-use planning* activities. In this respect, special consideration should be given to natural-hazard prone areas, as high-risk zones may be unsuitable for the siting of new installations.

The Addendum also addresses *transboundary cooperation*, advocating joint activities related to Natech prevention, preparedness, and response including natural-hazard identification, the drafting and communication of natural-hazard maps, the establishment of natural-hazard warning systems, emergency planning including mutual assistance, and the improvement of Natech risk-management methodologies and requirements.

The 1988 OECD Recommendation on the Polluter Pays Principle foresees an exception if accidents are caused by unforeseeable severe natural disasters. The Natech Addendum invites countries to reflect on how to interpret this provision in light of known natural hazards and whether operators could be considered liable for damage and pollution triggered by extreme natural events.

4.3.1.3 *Further OECD Activities*
In 2015, the OECD's Working Group on Chemical Accidents decided to include a second Natech project in its work programme 2017–20. The aim of this project is:

- To demonstrate examples of the implementation of the OECD Natech Addendum.
- To present good practice examples of Natech risk management.
- To identify and fill gaps in existing recommendations related to Natech risk management.
- To improve international cooperation in Natech risk management.

This Natech-II-Project may also investigate the possible links between Natech risk management and natural disaster management.

4.3.2 The UNEP APELL Program
The Awareness and Preparedness for Emergencies at Local Level (APELL) program was developed by the United Nations Environment Programme (UNEP) after a request from governments in response to major accidents which occurred in 1984–85, including the Bhopal accident, but also the Mexico City, and Basel accidents. UNEP has implemented the APELL program in more than 30 countries and about 80 communities since 1988, to raise awareness about hazards and risks, improve preparedness planning, and prepare integrated plans.

APELL is a program developed by UNEP in conjunction with governments and industry with the purpose of minimizing the occurrence and harmful effects of technological hazards and environmental emergencies. The strategy of the APELL approach is to identify and create awareness of risks in a community, to initiate measures for risk reduction and mitigation, and to develop coordinated preparedness between the industry, the local authorities, and the local population. It aims at helping communities prevent loss of life; damage to health, well-being, and livelihoods; minimize property damage; and protect the environment. APELL objectives are: (1) raising awareness, communicating and educating the community, and (2) improving emergency-preparedness planning, through a multistakeholder participatory approach involving industry, the communities, and local authorities.

APELL was originally developed to tackle risks arising from fixed industrial installations. It has also been adapted, to be applied in specific contexts, through the following guidance: APELL for Port Areas (1996), TransAPELL for Dangerous Goods Transport (2000), APELL for Mining (2001), and APELL for Coastal Settlements in tourism destinations (2008). This last adaptation of APELL included a multihazard approach, taking into consideration natural hazards and technological hazards that could affect a community.

In 2015, a second edition of the APELL handbook was developed, for community awareness and preparedness for technological hazards and environmental emergencies. Among the new elements, this version integrates lessons learned from implementation over the past 30 years. In the handbook, both natural and technological hazards are considered, and UNEP recognizes the increased focus on technological hazards arising directly as a result of the impacts of a natural hazard event. This new edition therefore covers Natech events explicitly.

Often, at community level, the same persons or institutions are responsible for prevention and preparedness of natural hazards and technological hazards. It is even more the case with the APELL methodology as it highlights the importance of *integrated emergency plans at community level*. In this context, integrated means a preparedness process that is articulated with existing plans, including at regional/provincial levels, but also a process that takes a holistic approach to consider all potential events affecting the community, and the relation between these events. The handbook mentions: "*An effective planning process will help communities prevent loss of life, damage to health, well-being and livelihoods, minimise property damage, and protect the environment. These same goals apply regardless of the nature of the environmental emergency, whether it is an industrial accident, a natural disaster or a combination of events, such as might occur following an earthquake or tsunami disaster or smaller scale events such as lightning storms.*"

APELL is considered to be applicable to any risk situation, given its flexible methodology. APELL works through a 10-element process which involves ten elements: Identify Participants and Establish their Roles, Evaluate Hazards and Risks faced by the Community; Review Existing Capabilities and Emergency Plans—Identify Gaps; Create the Vision of Success; Make Progress toward the

Vision of Success; Make Changes in Existing Emergency Plans and Integrate into Overall Community Preparedness Plan; Obtain Endorsement from Government Authorities; Implement Community Preparedness Plans through Communicating, Educating, and Training Community Members; Establish Procedures for Periodic Testing, Review and Updating of the Plans; Maintain APELL through Continuous Improvement. This process creates a multistakeholder dialog and covers five phases: (I) engaging stakeholders, (II) understanding hazards and risks, (III) preparedness planning, (IV) implementing and testing, and (V) maintaining APELL. The methodology is relevant to the community as it is based upon awareness of the specific hazards and risks present and uses existing strengths and relationships. Tools and assets in a community for preparedness and response are often similar for industrial, natural, or Natech accidents, but specific attention should be given in some cases. For instance, during phase II, in the context of an industrial or Natech accident, knowledge on chemical hazards and specific emergency-response equipment will be required. In addition, different tools can be required for risk assessment and risk reduction. During phase IV, scenarios for drills and exercises should be adapted to include cases of potential Natech scenarios.

Over the years, while implementing APELL, local communities became aware of the local risks and its impacts, and prepared to respond appropriately in the event of an accident, and a better coordination and preparedness of emergency services at local level has been witnessed. In the next implementation of APELL methodologies, further application including Natech events is expected and should be encouraged.

> [Disclaimer for any external contribution of UNEP staff (articles, book chapter, etc.): "The author is a staff member of the United Nations Environment Programme. The author alone is responsible for the views expressed in the publication and they do not necessarily represent the decisions or policies of the United Nations Environment Programme."]

4.3.3 Sendai Framework for Disaster Risk Reduction 2015–30

Global disaster trends show that urbanization, population growth, and climate change have led to increasing numbers of fatalities, injuries, and economic losses from disasters over the past decade. The Sendai Framework for Disaster Risk Reduction 2015–30, which was adopted by the United Nations Member States in Mar. 2015 in Sendai, Japan, is the umbrella for work on disaster risk management at the international level (UNISDR, 2015). It charts the global course in disaster risk reduction until 2030 by setting priorities for action and provides the basis for the sustainable development agenda.

The Sendai framework is a voluntary, nonbinding agreement that has the following aim:

> "The substantial reduction of disaster risk and losses in lives, livelihoods and health and in the economic, physical, social, cultural and environmental assets of persons, businesses, communities and countries."

This represents a policy shift at the global level from managing disasters to managing disaster risks, thereby emphasizing the importance of prevention and preparedness, and moving from reactive to more proactive approaches to prevent new and reduce existing disaster risks. The Sendai framework defines four priorities for action and seven global targets whose achievement will be measured using appropriate indicators.

Recognizing the disaster potential of technological and Natech (cascading) hazards following the Fukushima Natech accident in 2011, the scope of disaster risk reduction has been broadened significantly in the Sendai framework which follows an all-hazards approach. As such, it includes natural and man-made hazards, and related environmental, technological, and biological hazards and risks. The framework also calls for a substantial reduction of disaster damage to critical infrastructure and disruption to their services. Since a loss of lifelines can trigger or exacerbate the consequences of a technological or Natech accident, this target will also directly support Natech risk reduction.

References

BRD, 2015. Gesetz zum Schutz vor schädlichen Umwelteinwirkungen durch Luftverunreinigungen, Geräusche, Erschütterungen und ähnliche Vorgänge, Bundesrepublik Deutschland. https://www.gesetze-im-internet.de/bundesrecht/bimschg/gesamt.pdf (in German).

Bundesregierung, 2008. Deutsche Anpassungsstrategie an den Klimawandel, German Government, Berlin. http://www.bmub.bund.de/fileadmin/bmu-import/files/pdfs/allgemein/application/pdf/das_gesamt_bf.pdf (in German).

CalARP, 2014. Guidance for California Accidental Release Prevention (CalARP) Program Seismic Assessments, CalARP Program Seismic Guidance Committee. http://www.caloes.ca.gov/FireRescueSite/Documents/SGD%20LEPC%20I%20Approved%2003%2012%202014.pdf

CAO, 2008. Fire Service Law, Law No. 186 of 1948, Amendment Law No. 41 of 2008, Cabinet Office, Government of Japan. http://www8.cao.go.jp/kisei-kaikaku/oto/otodb/english/houseido/hou/lh_05070.html

CAO, 2012. Petroleum Refinery Complex, Etc. Disaster Prevention Law, Cabinet Office, Government of Japan. http://www8.cao.go.jp/kisei-kaikaku/oto/otodb/english/houseido/hou/lh_05080.html

Cruz, A.M., 2014. Reporte de Recomendaciones, Semana de intercambio de experiencias internacionales en análisis de riesgo accidental y vulnerabilidad asociada al transporte de hidrocarburos, PNUD-ECOPETROL, Bogotá, Colombia (in Spanish).

Cruz, A.M., Okada, N., 2008. Consideration of natural hazards in the design and risk management of industrial facilities. Nat. Hazards 44, 213.

EPA, 2015. Risk Management Plan (RMP) Rule, US Environmental Protection Agency. http://www.epa.gov/rmp

European Union, 1997. Council Directive 96/82/EC of 9 December 1996 on the control of major-accident hazards involving dangerous substances. Off. J. Eur. Commun. 10, 13.

European Union, 2000. Directive 2000/60/EC of the European Parliament and of the Council of 23 October 2000 establishing a framework for Community action in the field of water policy. Off. J. Eur. Commun. 327, 43.

European Union, 2003. Directive 2003/105/EC of the European Parliament and of the Council of 16 December 2003 amending Council Directive 96/82/EC. Off. J. Eur. Commun. 345, 97.

European Union, 2007. Directive 2007/60/EC of the European Parliament and of the Council of 23 October 2007 on the assessment and management of flood risks. Off. J. Eur. Union. 288, 27.

European Union, 2012. Directive 2012/18/EU of the European Parliament and of the Council of 4 July 2012 on the control of major-accident hazards involving dangerous substances, amending and subsequently repealing Council Directive 96/82/EC. Off. J. Eur. Union. 197, 1.

European Union, 2013. Directive 2013/30/EU of the European Parliament and of the Council of 12 June 2013 on safety of offshore oil and gas operations and amending Directive 2004/35/EC. Off. J. Eur. Union. 178, 66.

HPGSIJ, 2005. High Pressure Gas Safety Law, Law No. 204 of June 7, 1951, Amendment: Law No. 73 of 2005, High Pressure Gas Safety Institute of Japan. http://www.khk.or.jp/english/activities.html

IPCC, 2007. Climate Change 2007 Impacts, Adaptation and Vulnerability, Fourth Assessment Report, Intergovernmental Panel for Climate Change. Cambridge University Press, Cambridge, UK.

Köppke, K.-E., Sterger, O., Stock, M., 2013. Grundlagen für die Technische Regel für Anlagensicherheit (TRAS) 310: Vorkehrungen und Maßnahmen wegen der Gefahrenquellen Niederschläge und Hochwasser (Basis for the Technical Rule on Installation Safety TRAS 310). Dessau-Roßlau, Umweltbundesamt.

Krätzig, W., Andres, M., Niemann, H.-J., Köppke, K.-E., Stock, M., 2016. Technische Regel für Anlagensicherheit (TRAS) 320: Vorkehrungen und Maßnahmen wegen der Gefahrenquellen Wind, Schnee- und Eislasten (Technical Rule on Installation Safety TRAS 320). Dessau-Roßlau, Umweltbundesamt.

Krausmann, E., 2010. Analysis of Natech risk reduction in EU Member States using a questionnaire survey, Report EUR 24661 EN, European Union.

Krausmann, E., Baranzini, D., 2009. Natech Risk Reduction in OECD Member Countries, JRC Scientific and Technical Report 54120, European Communities, limited distribution.

Krausmann, E., Baranzini, D., 2012. Natech risk reduction in the European Union. J. Risk Res. 15 (8), 1027.

OECD, 2003. Guiding Principles for Chemical Accident Prevention, Preparedness and Response, second ed. OECD Series on Chemical Accidents No. 10, Paris.

OECD, 2013. Report of the Workshop on Natech Risk Management, OECD Series on Chemical Accidents No. 25, Paris.

OECD, 2015. Addendum No. 2 to the OECD Guiding Principles for Chemical Accident Prevention, Preparedness and Response (second ed.) to Address Natural Hazards Triggering Technological Accidents (Natechs), OECD Series on Chemical Accidents No. 27, Paris.

OSHA, 2012. Process safety management of highly hazardous chemicals, 29 CFR 1910.119. Occupational Safety & Health Administration, US Department of Labor. https://www.osha.gov/pls/oshaweb/owadisp.show_document?p_table=STANDARDS&p_id=9760

PAJ, 2011. Petroleum Industry in Japan: 2011, Petroleum Association of Japan, September 2011. http://www.paj.gr.jp/english/data/paj2011.pdf

UNISDR, 2015. Sendai Framework for Disaster Risk Reduction 2015–2030. http://www. unisdr.org/files/43291_sendaiframeworkfordrren.pdf

Warm, H., Köppke, K.-E., 2007. Schutz von neuen und bestehenden Anlagen und Betriebsbereichen gegen natürliche, umgebungsbedingte Gefahrenquellen. Dessau-Roßlau, Umweltbundesamt.

Natural Hazard Characterization

G. Lanzano*, A. Basco**, A.M. Pellegrino†, E. Salzano‡
*National Institute of Geophysics and Volcanology (INGV), Milan, Italy; **AMRA, Analysis and Monitoring of Environmental Risk, Naples, Italy; †Engineering Department, University of Ferrara, Ferrara, Italy; ‡Department of Civil, Chemical, Environmental, and Materials Engineering, University of Bologna, Bologna, Italy

Major natural events must be characterized in terms of their hazard, that is, their ability to cause significant harm. Although there have been attempts to predict the occurrence of natural events, uncertainties still exist with respect to large-scale natural hazards due to their complexity. Furthermore, for Natech risk analysis, a simplified event characterization is needed. This chapter focuses on the characterization of selected natural hazards which were found to be relevant with respect to impacts at industrial installations, that is, earthquakes, floods, and, due to recent events in Japan, tsunamis.

5.1 INTRODUCTION

Natural events can affect the integrity of industrial installations and cause damage, business interruption, and subsequent economic losses, but they can also induce major accidents due to the release of energy or materials into the atmosphere (Salzano et al., 2013; Krausmann et al., 2011). In this framework, earthquakes, floods, lightning, tsunamis, and storms (hurricane, tornado), as well as other natural-hazard phenomena, such as intense rain or extreme temperatures, must be characterized in terms of their hazard, that is, their ability to cause harm in significant amounts, and to possibly overwhelm the capacity of industrial and public emergency-response systems. These issues are a matter of concern for public authorities and the population. Indeed, since the beginning of human civilization, there have been attempts to predict the occurrence of natural events. Therefore, the body of scientific literature, guidelines, standards, and historical databases on natural events, and more specifically on the hazard they represent, is large. For each of the cited events, the knowledge in the hands of scientists and practitioners is nowadays sufficiently large for the development of efficient measures and systems to mitigate the associated risks.

Despite these observations, and in the light of Natech risk reduction, public authorities, industry managers, and risk analysts often need a simplified characterization and representation of natural phenomena. This allows a proactive development and adoption, in the early design phase, of structural and organizational protection measures. To this end, a characterization with only 1 or 2 degrees of freedom for each natural phenomenon of concern is needed.

This requirement was implemented by Cornell (1968), in his pioneering work on earthquake and reliability engineering, who along with Luis Esteva developed the field of Probabilistic Seismic Hazard Analysis (PSHA) for earthquakes which is nowadays used worldwide (Baker, 2008). Following his general methodology, the hazard value (or hazard curve) H for the given natural event (nat) may then be expressed as the cumulative probability of exceeding a threshold intensity value λ_{nat}, over the time interval T as:

$$H_{nat}(\lambda_{nat}, T) = \phi(\lambda_{nat} > \alpha \,|\, T) \tag{5.1}$$

where α is the reference intensity value and the function ϕ is evaluated by the Poisson law for the global annual rate of occurrence R_{tot} of rare events which is independent of time, that is:

$$\phi = 1 - e^{-R_{tot}T} \tag{5.2}$$

The variable R_{tot} is typically represented by the summation over the standard, normal, cumulative probability distribution functions of each natural- event scenario, and it includes information on the mean and the standard deviation:

$$R_{tot}(\lambda_{nat} > \alpha) = \sum_{scenario} \frac{1}{\sigma\sqrt{2\pi}} \int_{\lambda_{nat}}^{\infty} e^{-(\lambda_{nat}-\mu)^2/2\sigma} d\lambda_{nat} \tag{5.3}$$

Where industrial equipment is concerned, hazard curves could be calculated for a time T that corresponds to the Technical Service Life (TSL) of equipment. In some cases, designers consider the Functional Service Life (FSL) which is lower than the TSL (ISO, 2000). However, the first option satisfies the need for conservative choices to be on the safe side. Eventually, the following expression may be adopted:

$$H_{nat}(\lambda_{nat}, TSL) = \phi(\lambda_{nat} > \alpha \,|\, TSL) \tag{5.4}$$

Often, 50 years is assumed as the TSL. This choice is not only technical. Indeed, the 10% exceedance probability of occurrence of λ_{nat} in 50 years is associated with the mean earthquake recurrence interval of 475 years. Hence, a 475-year mean recurrence interval has tended to become a widely used benchmark value for acceptable risk.

The annual exceedance probability of occurrence, R_{tot}, is often calculated using a Weibull equation in terms of the number n of historical observations corresponding to a given intensity or magnitude, expressed as m, and it represents per se a measure of the hazard:

$$R_{tot}(\lambda_{nat} > \alpha) = \frac{n+1}{m} \tag{5.5}$$

From this equation, the return period (or the recurrence interval) T_R, which is the average number of years between the occurrence of two events of equal (or greater) magnitude, can be derived. It is the reciprocal of R_{tot}.

These issues will be extensively discussed for the most important natural events of concern in this book: earthquake, tsunami, and flooding. For each of these events, this chapter will evaluate the main hazard intensity parameters and the related limitations and uncertainties. A similar approach can be adopted for any other natural-hazard phenomenon.

5.2 PREDICTION AND MEASUREMENT

5.2.1 Earthquake

Earthquakes are seismic movements of the ground that are mainly caused by tectonic activities. The interaction between earthquakes and equipment can result in large damage when hazardous industrial installations are involved (Campedel et al., 2008). Clearly, Natech risk reduction needs a multidisciplinary effort that involves (1) the definition of the occurrence probability of a given earthquake intensity, (2) the structural analysis of the equipment interaction with the seismic forces, (3) the analysis of the specific response of the industrial process after structural damage of the equipment and its supports has occurred, and finally (4) the analysis of risks in global terms, following classical methodologies.

5.2.1.1 Hazard Parameters of Concern

With specific reference to earthquakes and given a hazardous industrial site of interest, ground motion is the main variable to take into account. More specifically, the measured ground motion refers to seismic waves radiating from the earthquake focus to the hazardous site and is related to the earthquake source, the path of the seismic wave from the source to the site, and to the specific geomorphologic characteristics of the site. Moreover, the earthquake characteristics include energy, frequency contents, phases, and many other variables that can affect the structural response of buildings and other structures.

Currently, the problem of defining effective and reliable predictors for the seismic response behavior of structures is one of the main topics of earthquake engineering. Empirical vulnerability analyses are often carried out in terms of peak ground acceleration, PGA (or alternatively in terms of peak ground velocity, PGV), as this parameter is relatively easy to infer by earthquake intensity conversion (Panico et al., 2016; Lanzano et al., 2013, 2014a; Salzano et al., 2003, 2009). Furthermore, several databases on historical damage due to earthquakes are usually related to the PGA of the earthquake [e.g., the pipeline damage database provided by Lanzano et al. (2015)]. Moreover, the calculation of this parameter from typical earthquake magnitude metrics (e.g., the Modified Mercalli scale or the Richter scale) is straightforward.

Local and national authorities usually provide tools for PSHA (Bommer, 2002; Cornell, 1968) both in Europe and in the USA. Hence, the exceedance probability of PGA occurrence calculated over 1 year or on a 50-year basis is nowadays available. The exceedance probability curve is the general reference function for structural-design purposes. As a matter of fact, seismic loads are usually determined from the maximum PGA of an earthquake at the site of interest over a given time period (the TSL or other reference time intervals given by standards or legislation).

The PGA data from the Global Seismic Hazard Assessment Project (GSHAP), expressed as percentage of the earth's gravity, g, is typically used as a reference for the earthquake hazard map. The GSHAP has produced a homogeneous seismic hazard map for horizontal PGAs representative of stiff site conditions, with a 10% chance of exceedance in 50 years. In addition, hazard curves and classifications can be found in local information systems. In this context, the US Geological Survey is a good source of seismic hazard maps (http://www.usgs.gov).

A discussion of the secondary natural hazards related to earthquakes, such as soil liquefaction and ground displacement, is outside the scope of this chapter, although these phenomena can produce significant damage to industrial equipment (Lanzano et al., 2014b; Krausmann et al., 2010).

5.2.1.2 Probabilistic Seismic Hazard Analysis

PSHA is a probability-based framework to evaluate the seismic hazard (Baker, 2008; Kramer, 1996). According to Baker (2008), PSHA is composed of the five steps described in Fig. 5.1. This approach aims to identify the annual rate of exceeding a given ground-motion intensity by considering all possible earthquakes and the associated ground motions together with their occurrence probabilities, thereby avoiding the definition of a worst-case ground-motion intensity which is not without difficulty.

For Natech risk analysis it is important to know the ground shaking at a hazardous site. In order to predict the ground motion, the distance distribution from earthquakes to the site of interest needs to be modeled. Baker (2008) notes that for a given earthquake source, an equal occurrence probability is generally assumed at any location on the fault. If locations are uniformly distributed, the distribution of source-to-site distances by using the geometry of the source is straightforward.

For any earthquake, if IM is the ground-motion intensity measure of interest (such as PGA), the natural logarithm of IM is normally distributed. Hence, at any distance r from the earthquake source with magnitude m, the probability of exceeding any PGA level x can be evaluated by using the corresponding mean and standard deviation σ:

$$P\left(IM > x | m, r\right) = 1 - \Phi\left(\frac{\ln x - \overline{\ln IM}}{\sigma_{\ln IM}}\right) \quad (5.6)$$

where Φ is the standard normal cumulative distribution function (Baker, 2008). Considering the number of possible sources, $n_{sources}$, we can then integrate over all considered magnitudes and distances in order to obtain $\lambda\,(IM > x)$, that is, the rate of exceeding IM:

$$\lambda(IM > x) = \sum_{i=1}^{n_{sources}} \lambda(M_i > m_{min}) \int_{m_{min}}^{m_{max}} \int_{0}^{r_{max}} P(IM > x | m, r)\, f_{M_i}(m)\, f_{R_i}(r)\, dr\, dm \quad (5.7)$$

where $\lambda\,(M_i > m_{min})$ is the rate of occurrence of earthquakes greater than m_{min} from the source i, and $f_M(m)$ and $f_R(r)$ are the probability density functions (PDFs) for magnitude and distance.

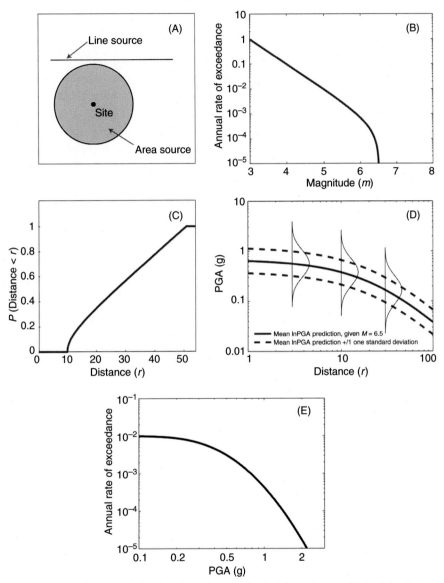

FIGURE 5.1 Schematic of the Five Basic Steps in Probabilistic Seismic Hazard Analysis According to Baker (2008)

(A) Identify earthquake sources, (B) characterize the distribution of earthquake magnitudes from each source, (C) characterize the distribution of source-to-site distances from each source, (D) predict the resulting distribution of ground motion intensity, (E) combine information from A–D to calculate the annual rate of exceeding a given ground motion intensity.

Occasionally, the results of a PSHA are expressed in terms of return periods of exceedance, that is, the reciprocal of the rate of occurrence, or as probabilities of exceeding a given ground-motion intensity within a specific time window for a given rate of exceedance. The latter is calculated under the assumption that the probability distribution of time between earthquakes is Poissonian. The probability of observing at least one event in a period of time t is therefore equal to (Baker, 2008):

$$P_{(\text{at least one event in time } t)} = 1 - e^{-\lambda t} \approx \lambda t, \tag{5.8}$$

where λ is the rate of event occurrence as defined earlier. For an exhaustive discussion of PSHA the reader is referred to Baker (2008) and Kramer (1996).

For design applications, there has emerged a convention to consider the probability of a ground-motion level within a given design life (T_L) of a structure in question [$t = T_L$ in Eq. (5.8)]. Hence for a given probability and design life, the seismic hazard can also be expressed in terms of the return period T_R given by:

$$T_R = \frac{-T_L}{\ln(1-P)} \tag{5.9}$$

It is common to see seismic hazard defined in terms of P and T_L in the current codes and provisions, for example, in Eurocode 8 (CEN, 2004).

For some critical structures, such as nuclear power plants, the design ground motion may be more commonly expressed as an annual probability or frequency of being exceeded (i.e., $T_L = 1$ year). Under International Atomic Energy Association regulations (IAEA, 2003), a typical design criterion for critical elements of a nuclear power station is a ground motion with an annual probability of being exceeded of 10^{-4}. Using Eqs. (5.8) and (5.9), this corresponds to approximately a 1% probability of being exceeded in 100 years (or a 0.5% probability of being exceeded in 50 years).

More generally, the performance-based seismic design (PBSD) formalizes the approach of citing multiple objectives for structures to withstand minor or more frequent levels of shaking with only nonstructural damage, while also ensuring life-safety and the avoidance of collapse under severe shaking (ATC, 1978). These objectives define the limit states, which describe the maximum extent of damage expected to the structure for a given level of ground motion.

Although there are different definitions of limit states, a 475-year return period (corresponding to $P = 10\%$ in $T_L = 50$ years) is commonly adopted as a basis for ensuring "life-safety." However, several codes have recently begun to adopt 2475 years (corresponding to $P = 2\%$ in $T_L = 50$ years) as the return period for the no-collapse criterion, even though it is subsequently rescaled to incorporate an assumed inherent margin of safety against collapse (NBCC, 2005; NEHRP, 2003). Longer return periods may be considered for critical structures, but this kind of analysis would require that uncertainties be treated carefully.

The 2009 revision to the NEHRP Provisions introduces a new conceptual approach for the definition of the input seismic action (NEHRP, 2009). The seismic input (maximum considered earthquake) is modified by a risk coefficient (for both short and long periods) which is derived from a probabilistic formulation of the

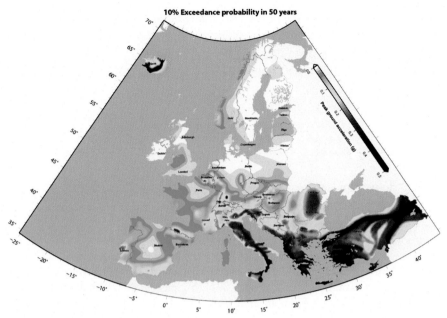

FIGURE 5.2 European Seismic Hazard Map in Terms of Exceeding a Peak Ground Acceleration (PGA) With a Probability of 10% in 50 Years

likelihood of collapse (Luco et al., 2007). These modifications change the definition of seismic input to ensure a more uniform level of collapse prevention.

For most applications, the hazard is described in terms of a single parameter, that is, the value of the reference PGA on type A ground, which corresponds to rock or other rock-like geological formations. As an example, Fig. 5.2 shows the European Seismic Hazard Map (ESHM) which illustrates the probability to exceed a level of ground shaking in terms of the PGA in a 50-year period. The illustrated levels of shaking are expected to be exceeded with a 10% probability in 50 years, which corresponds to a return period T_R of 475 years.

European legislation (CEN, 2004) prescribes the use of zones for which the reference PGA hazard on a "rock" site (a_g) is assumed uniform. Many seismic codes are moving away from this particular practice, choosing instead to define the hazard directly for the site under consideration (NEHRP, 2003, 2009; NTC, 2008; NBCC, 2005), or allowing for interpolation between contour levels of uniform hazard.

As an example and to give an order of size of the acceleration levels in Europe, for a medium-high seismicity area like Italy four seismic zones are defined (OPCM, 2003) according to the value of the maximum horizontal peak ground acceleration a_g whose probability of exceedance is 10% in 50 years (Table 5.1). Subsequently, in the new Italian Building Code (NTC, 2008) the Italian Hazard Map (called MPS04) was defined for each single location and is therefore site specific.

Table 5.1 Seismic Zonation in Italy According to the OPCM (2003)

Seismic Zone	Ground Acceleration (g) With Probability of Exceedance Equal to 10% in 50 Years (a_g)	Acceleration (g) of the Elastic Response Spectrum (a_g) at period $T = 0$
1	>0.25	0.35
2	>0.15–0.25	0.25
3	0.05–0.15	0.15
4	<0.05	0.05

5.2.2 Tsunami

A tsunami (lit. "harbor wave") is a series of long-period water waves caused by the displacement of a large volume of water generated by undersea earthquakes, volcanic eruptions, aerial landslides, and other disturbances above or below water. The occurrence of tsunamis can cause major (direct and indirect) losses in terms of human lives and infrastructure damage as seen recently in Japan (2011), Chile (2010), and on the coastlines of the Indian Ocean (2004) (Suppasri et al., 2013; Mas et al., 2012; Koshimura et al., 2009). From an industrial-safety point of view, port structures are particularly vulnerable.

Based on historical tsunami observations, the vast majority of tsunamis are induced by earthquakes. The resulting waves have small amplitude (the wave height above the normal sea surface), and a very long wavelength (often hundreds of kilometers), whereas normal ocean waves have a height of roughly 2 m and a wavelength of only 100–200 m. Tsunami waves can travel at speeds of over 800 km/h in the open sea. Due to the large wavelength, the wave takes 20–30 min to complete a cycle and has an amplitude of less than 1 m. For this reason, tsunamis are difficult to detect in deep water.

Like other types of waves, tsunami waves have a positive (ridge) and negative peak (trough). If the ridge arrives first, a huge breaking wave or sudden and quick flooding on land occurs. The resulting temporary rise in sea level is called run-up. Run-up is measured in meters above a reference sea level. If the tsunami trough arrives first, the shoreline recedes as the water is drawn back. A large tsunami may exhibit multiple waves arriving over a period of hours, with significant time between the wave crests. It is interesting to note that the first wave to reach the shore may not have the highest run-up (Nelson, 2012).

Tsunamis can cause damage via two mechanisms: (1) the force of a wall of water traveling at high speed slamming into coastlines and structures, and (2) the drag forces of a large water volume that recedes from the land, thereby carrying a large amount of debris with it. This latter phenomenon can occur even with waves that do not appear to be large. From an engineering point of view, the tsunami action needs to be evaluated in terms of loading on structures, and proper intensity parameters must be considered. In the common design practice, the relevant parameters are the maximum water height, h_w and the maximum water velocity, v_w.

Probabilistic tsunami hazard analysis (PTHA) is similar to the widely used PSHA for earthquakes. The basic approach is to combine the rate at which tsunamis are generated with the distribution of amplitudes that are expected to occur at the site for a given tsunami. The probabilistic tsunami hazard from earthquakes is given by (PGEC, 2010; Rikitake and Aida, 1988):

$$\nu_{EQK}\left(W_{tsu} > z\right) = \sum_{i=1}^{N_{FLT}} N_i\left(M_{min}\right) \int_m \int_{Loc} f_{m_i}(M) f_{Loc_i}(Loc) P\left(W_{tsu} > z | M, Loc\right) dMdLoc \quad (5.10)$$

where $\nu_{EQK}\left(W_{tsu} > z\right)$ is the annual rate of tsunami wave heights exceeding z, N_{FLT} is the number of tsunamigenic fault sources, $N_i\left(M_{min}\right)$ is the rate of earthquakes with magnitude greater than M_{min} for source i, f_m and f_{Loc} are PDFs for the magnitude and rupture location, and $P(W_{tsu} > z | M, Loc)$ is the conditional probability of the tsunami wave height, W_{tsu}, exceeding the test value z.

Assuming the tsunami wave heights are log-normally distributed, the conditional probability of exceeding wave height z is given by (PGEC, 2010):

$$P(W_{tsu} > z | M, Loc) = 1 - \Phi\left(\frac{\ln(z) - \ln(\hat{W}_{tsu}(M, Loc))}{\sigma_{EQK}}\right) \quad (5.11)$$

where $\hat{W}_{tsu}(M, Loc)$ is the median wave height, σ_{EQK} is the aleatory variability of the tsunami wave height from earthquakes (e.g., standard deviation) in natural logarithm units, and Φ is the cumulative normal distribution.

If only a small number of representative scenarios (magnitude and location) are considered, then the tsunami hazard from earthquakes simplifies to (PGEC, 2010):

$$\nu_{EQK}(W_{tsu} > z) = \sum_{i=1}^{N_{FLT}} \sum_{j=1}^{NS_i} rate_{ij} P(W_{tsu} > z | M_{ij}, Loc_{ij}) \quad (5.12)$$

where $rate_{ij}$ is the rate of occurrence of scenario j from source i.

The tsunami wave heights are calculated using numerical simulations. A site-specific PTHA involves the production of a full source-to-site numerical tsunami simulation on a high-resolution digital elevation model (DEM) for each considered potential source scenario. As a result, the computational cost of site-specific PTHA would generally be almost unaffordable in practice. Hence, specific strategies are being developed for reducing the computational burden (Geist and Lynett, 2014). These strategies are typically based on crude approximation methods extrapolating inland the offshore wave heights, and/or using an oversimplification of the seismic source variability and applying a cruder selection of the relevant seismic sources (Thio et al., 2010). These procedures are therefore affected by very large epistemic uncertainties.

A novel methodology to reduce the computational cost associated with a site-specific tsunami hazard assessment for earthquake-induced tsunamis has recently been presented by Lorito et al. (2015). It allows the performance of high-resolution inundation simulations on realistic topobathymetry only for the relevant seismic sources.

Table 5.2 Flood Hazard Classification Based on Observations Over a Period of 15 Years (ESPON, 2006)

Number of Observed Major River Floods at NUTS3 Level	Hazard Classes	Definition
0	1	Very low hazard
1	2	Low hazard
>1 –≤2	3	Moderate hazard
>2–≤3	4	High hazard
>3	5	Very high hazard

5.2.3 Floods

The US Federal Emergency Management Agency (FEMA) defines flooding as a general and temporary condition of partial or complete inundation of normally dry land areas from the overflow of inland or tidal waters, from the unusual and rapid accumulation or runoff of surface waters from any source, and from mudflows. Most floods fall into three major categories: riverine, coastal, and shallow flooding. Alluvial fan flooding is another type of flooding more common in mountainous areas (Pellegrino et al., 2015, 2010).

Flood hazard maps are available in many regions in Europe and the United States. Often, those maps report the number of observed floods with specific magnitude in the areas of concern over a given time interval, thus following a Weibull analysis. Table 5.2 shows a proposal for a flood hazard classification as reported by the European Spatial Planning Observation Network (ESPON, 2006).

The identification of flood-prone areas requires the collection and analysis of historical flood data, the availability of accurate digital elevation data, water discharge data, and stream cross-sections located throughout the watershed (Baban, 2014). In Europe, this data is available only for certain case-study areas. So far, the mapping of flood-prone areas in Europe does not follow a consistent approach, and there are several approaches in different catchment areas or riverbeds.

For the purpose of Natech risk analysis, the area affected by flooding, or alternatively, the intensity of the phenomenon, are identified by the maximum water depth (h_w) expected at the hazardous industrial site and/or the maximum expected water velocity (v_w). These two parameters are highly dependent on the flood scenario considered. Hence, different return times are usually assessed for each specific flood scenario.

In several European countries, flood hazard maps showing the maximum expected water depth and velocity given the return time of the flood event are available. Three categories of water impact are typically defined. With respect to *water velocity* these are: (1) slow submersion (negligible water velocity), (2) low-speed wave (water velocity ≤ 1 m/s), and (3) high-speed wave (water velocity > 1 m/s). Concerning *water height* the categories are: (1) low height (0.5 m, no damage expected),

Table 5.3 Flood Hazard Classification Based on Water Depth and Velocity

Hazard Index	Hazard Classification	Water Depth (m/s)	Water Velocity (m/s)
1	Very low	≤0.5	≤0.2
2	Low	>0.5–1	>0.2–0.5
3	Moderate	>1–1.5	>0.5–1.0
4	High	>1.5	>1.0

Table 5.4 Flood Hazard Classification Based on Number of Floods Observed in 50 Years

Hazard Index	Hazard Classification	Number of Observed Floods (Year^{-1})
1	Very low	0
2	Low	1–3
3	Moderate	4–6
4	High	>7

Adapted by the European Spatial Planning Observation Network (ESPON, 2006)

(2) intermediate height (1 m/s, damage expected), and (3) high water height (1.5 m, extensive damage expected).

Table 5.3 shows a hazard classification that is based on maximum water depth and velocity. If both values are available, those leading to the worst-case classification should be adopted. Naturally, such a hazard characterization is dependent on the time of return selected for the reference flood events.

If data concerning maximum water depth and water velocity are not available, a general natural hazard index with specific reference to Natech risks can be obtained from historical data. The approach originally developed by the European Spatial Planning Observation Network (ESPON, 2006) can be adopted to obtain a general hazard characterization based on a 50-year observational period. The resulting hazard matrix is shown in Table 5.4.

If flood maps are available that only show the maximum extent of the flooded area for a given return time, a simplified estimation of the maximum water depth and velocity can be obtained as follows:

- The maximum water depth may be assumed as the difference between the height of the soil at the boundary of the flooded zone and the mean height of the site of interest.
- The mean velocity of a gravity-driven flow in rough open channels and rivers can be estimated using Manning's empirical formula:

$$v = \frac{1}{n} s^{1/2} R^{2/3}$$

(5.13)

where v (m/s) is the mean velocity of the flow, s (m/m) is the slope of the channel if the water depth is constant, R (m) is the hydraulic radius of the cross-section of the channel (defined as the area of the cross-section of the channel divided by the length of the wetted perimeter, which is easily determined assuming a simplified trapezoidal shape of the channel), and n is a roughness coefficient that can be related through standard values to the river characteristics (e.g., $n = 0.030$ for clean and straight rivers, $n = 0.035$ for major rivers, and $n = 0.040$ for sluggish rivers with deep pools).

The FEMA Flood Map Service Center (MSC) is the official public source for flood hazard information produced in support of the National Flood Insurance Program (NFIP) in the USA (https://msc.fema.gov/). MSC produces official flood maps and gives access to a range of other flood hazard products. Generally, three main approaches are taken into consideration to address the risks due to flooding:

- Statistical studies to determine the probability and frequency of high discharges of streams that cause flooding.
- Analytical models and maps to determine the extent of possible flooding when it occurs in the future.
- Monitoring storms and snow levels to provide short-term flood prediction, since the main causes of flooding are abnormal amounts of rainfall and sudden thawing of snow or ice.

5.2.3.1 Probability and Frequency of Flooding

If data on stream discharge is available over an extended period of time, it is possible to determine the flood frequencies for any given stream. Starting from historical observations, statistical analysis can be used to determine how often a given discharge or stage of a river is expected. This allows the definition of a return period or recurrence interval and the probability of a given discharge in the stream for any year. The yearly maximum discharge of a stream from one gauging station over a sufficiently long period of time is needed for this analysis.

As a first step in the determination of the recurrence interval, the yearly discharge values are ranked (Nelson, 2015). Each discharge is associated with a rank, m, with $m = 1$ assigned to the maximum discharge over the years of record and $m = n$ attributed to the smallest discharge whose rank is equal to the number of years over which there is a record, n. Using the following Weibull equation, the number of years of record, n, and the rank for each peak discharge are then used to calculate the recurrence interval, R:

$$R = \frac{n+1}{m} \tag{5.14}$$

Knowing the recurrence interval and the yearly discharge, these two quantities can be combined in a plot that allows the determination of the expected peak discharge for floods with specific return periods. An example of such a plot is shown in Fig. 5.3 for the Red River of the North gauging station at Fargo, North Dakota.

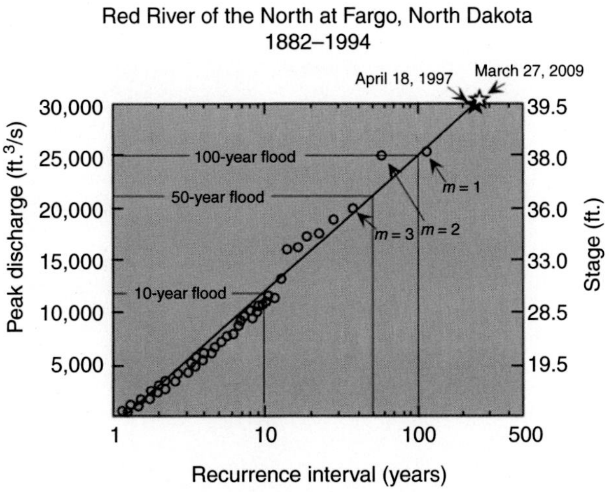

FIGURE 5.3 **Frequency of Flooding: Relation Between the Peak Discharge for Each Year Versus Recurrence Interval of the Red River in Fargo, North Dakota**

From E.M. Baer (2007).

The probability, P_e, of a specific stream discharge occurring in any year can be calculated using the inverse of Eq. (5.14). P_e is also called the annual exceedance probability (Nelson, 2015):

$$P_e = \frac{m}{n+1} \tag{5.15}$$

The probability that one or more floods occurring during any period exceed a given flood severity can be calculated using the following equation:

$$P_t = 1 - (1 - P_e) \cdot n \tag{5.16}$$

where P_t is the probability of occurrence over the entire time period, n.

5.2.3.2 Flood Maps

Flood hazard maps illustrate the areas susceptible to flooding when a river exceeds its banks due to different discharge scenarios. Coupled with topographic information and supported by satellite images and aerial photography of past flood events, these maps can be created using historical data on river stages and discharge levels of previous floods (Nelson, 2015). Fig. 5.4 shows a hazard map for a hypothetical 10–20-year, a 100-year, and a 200-year flood for a region in Germany crossed by the Elbe River. While flood hazard maps contain information on the magnitude and likelihood of a flood event, flood *risk* maps also include information on the potential consequences of flooding.

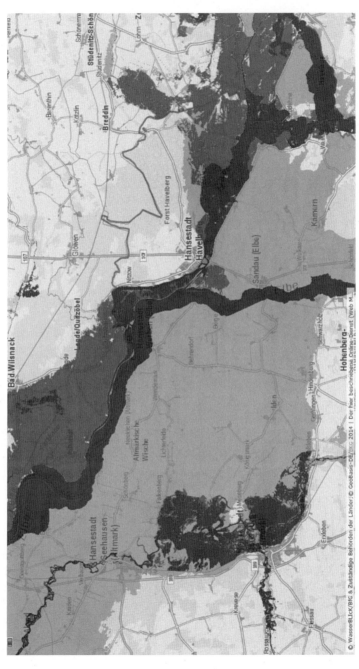

FIGURE 5.4 Hazard Map for Hypothetical 10–20-Year, 100-Year, and 200-Year Inundation Scenarios for the River Elbe

Bundesanstalt für Gewässerkunde, Germany, http://geoportal.bafg.de/mapapps/resources/apps/HWRMRL-DE/index.html?lang=de

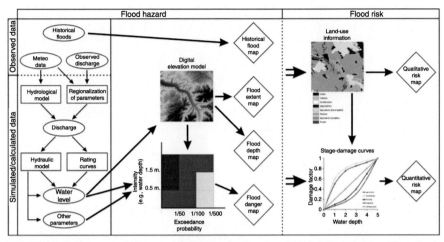

FIGURE 5.5 Conceptual Framework for Flood Hazard and Risk Mapping for a Hypothetical Case

Courtesy of de Moel et al. (2009).

There are different methods to quantify flood hazards and risks which result in different types of flood maps as illustrated in Fig. 5.5. Flood hazards can be evaluated using methods of lower or higher complexity which depend on the available data, resources, and time. Nevertheless, de Moel et al. (2009) indicate that the conceptual framework behind the calculation of flood hazards is general and in principle follows three steps:

1. Estimation of the discharge levels for specific return periods. Most commonly, hydrological models are used to calculate discharges. These models require spatially explicit and comprehensive knowledge on meteorological conditions, soil, and land cover. Alternatively, discharge levels can be determined by frequency analyses of discharge records and fitting of extreme-value distributions, or in case this information is not available, from precipitation records using runoff coefficients.
2. Translation of discharge levels into water levels once discharges and their associated return periods have been derived. This is usually accomplished using so-called rating (stage-discharge) curves or with 1D or 2D hydrodynamic models.
3. Determination of the inundated area (and—if possible—also of the flood depth) by combining water levels with a DEM. This procedure yields a flood map that shows either flood extent or depth.

Flood extent maps are the most common flood hazard maps. They show the inundated areas for a specific scenario which can either be a historical or a hypothetical flood event with a specific return period, for example, once in 50 or 100 years (Fig. 5.6B). When the extent of a flood is known for specific return periods and a

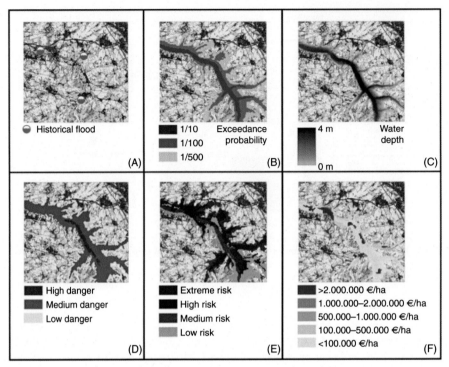

FIGURE 5.6 Different Flood Map Types Based on a Hypothetical Case

(A) Historical flood map, (B) flood extent map, (C) flood depth map, (D) flood danger map, (E) qualitative risk map, (F) quantitative risk (damage) map.

Courtesy of de Moel et al. (2009).

DEM is available, the flood depths can be easily derived. This process results in a flood depth map (Fig. 5.6C).

Generally, in flood hazard maps the inundation area and the related water depths are normally considered the most important parameters for estimating the flood's adverse consequences. However, other parameters, such as water velocity, the duration of the flood, or the rate of water rise can also be crucial depending on the circumstances of the flood. In particular with respect to Natech risks, the water velocity can be a determining factor as the vulnerability of hazardous equipment toward still water or fast water flows differs significantly.

5.2.3.3 Flood Forecasting and EU Floods Directive

Flood forecasting allows the alerting of authorities and the population of imminent flood conditions so that they can take appropriate protective actions. The most accurate flood forecasts use long time-series of historical data that relates stream flows to measured past rainfall events. A coupling procedure of historical information with real-time observations is needed in order to make the most accurate flood forecasts (Pellegrino et al., 2010, 2015) together with radar estimates of rainfall and general weather-forecasting techniques. The intensity and height of a flood can be predicted with good accuracy and significant lead-time if high-quality data is available. Flood forecasts typically provide parameters like the maximum expected water level and the likely time of its arrival at specific locations along a waterway. There are several regulations, initiatives, and programs in place that support flood forecasting and general risk reduction from flooding. Some selected examples are provided as follows.

In many countries, urban areas prone to flooding are protected against a 100-year flood, that is, a flood that has a probability of around 63% of occurring in any 100-year period of time. The United States National Weather Service (NWS) Northeast River Forecast Center (NERFC) (http://www.weather.gov/nerfc/), which is a part of the US National Oceanic and Atmospheric Administration (NOAA), assumes for flood forecasting in urban areas that it takes at least 1 in. (25 mm) of rainfall in about 1 h to start significant ponding of water on impermeable surfaces. Many River Forecast Centers of the NWS routinely issue Flash Flood Guidance and Headwater Guidance, which indicate the amount of rainfall that would need to fall in a short period of time to cause flooding (http://www.srh.noaa.gov/rfcshare/ffg.php).

In addition, the United States follows an integrated approach to producing real-time hydrologic forecasts which are available from the NWS (http://water.weather.gov/ahps/). This approach uses, for example, data on real-time, recent and past streamflow conditions from the U.S. Geological Survey [http://waterwatch.usgs.gov/, different community collaborative observing networks (http://www.cocorahs.org/)] and automated weather sensors, the NOAA National Operational Hydrologic Remote Sensing Center (http://www.nohrsc.noaa.gov/) etc. combined with quantitative precipitation forecasts (QPF) of expected rainfall and/or snow melt. This forecasting network also includes various hydroelectric companies.

The Global Flood Monitoring System (GFMS) is an experimental system funded by NASA which maps flood conditions quasi worldwide (http://flood.umd.edu). Users anywhere in the world can use GFMS to determine when floods might occur in their area. GFMS uses precipitation data from NASA's Global Precipitation Measurement (GPM) mission, a system of Earth observing satellites. Rainfall data from GPM is combined with a land surface model to determine how much water is soaking into the ground, and how much water is flowing into streamflow. Users can view statistics for rainfall, streamflow, water depth, and flooding every 3 h, at each 12-km grid point on a global map. Forecasts for these parameters are 5 days into the future. The resolution of the produced inundation maps is 1 km.

The FloodList project reports on all major flood events occurring to raise awareness of flood risks and the associated potential severe consequences (http://floodlist.com). FloodList has become an authoritative source of up-to-date information on flood events.

In response to several recent extreme flood events, the European Floods Directive on the assessment and management of flood risks requires EU Member States to develop Flood Risk Management Plans (European Union, 2007). These plans need to set appropriate objectives for the management of flood risk and reduce potential adverse consequences on a number of risk receptors. The Directive prescribes a three-step procedure to achieve its objectives:

- Preliminary Flood Risk Assessment: The Floods Directive requires Member States to engage their government departments, agencies, and other bodies to draw up a Preliminary Flood Risk Assessment. This assessment has to consider impacts on human health and life, the environment, cultural heritage, and economic activity.
- Risk Assessment: The information in this assessment will be used to identify the areas at significant risk which will then be modeled in order to produce flood hazard and risk maps. These maps include detail on the flood extent, depth, and level for three risk scenarios (high, medium, and low probability).
- Flood Risk Management Plans: These plans are meant to indicate to policy makers, developers, and the public the nature of the risk and the measures proposed to manage these risks. However, they are not formally binding (e.g., with respect to land-use planning). The Floods Directive prescribes an active involvement of all interested stakeholders in the process. The management plans are to focus on prevention, protection, and preparedness. Also, flood risk management plans shall take into account the relevant environmental objectives of Article 4 of the EU Water Framework Directive (European Union, 2000).

The Global Flood Awareness System (GloFAS, http://www.globalfloods.eu) was jointly developed by the European Commission and the European Centre for Medium-Range Weather Forecasts (ECMWF) in the United Kingdom. The system, which is independent of administrative and political boundaries, links state-of-the-art weather forecasts with a hydrological model. Since GloFAS has a continental scale set-up it provides downstream countries with information on upstream river conditions as well as continental and global overviews.

5.3 LIMITATIONS, UNCERTAINTIES, AND FUTURE IMPACTS OF CLIMATE CHANGE

Other natural phenomena could be included in the list of events discussed earlier. Many are strongly affected, either in terms of their frequency or intensity, by climate change (IPCC, 2007). Severe weather phenomena, such as heat and cold waves, tornadoes, cyclones (typhoons or hurricanes) but also intense rainfall, water sprouts, and extreme winds can cause major Natech events as discussed in Chapter 3. These phenomena have been loosely analyzed and only few detailed studies are available on the vulnerability of industrial equipment toward these natural hazards.

Finally, it is important to note that uncertainties in the assessment of natural hazards are often caused by a lack of knowledge or understanding of the underlying physical processes. In case of large-scale natural events, there is insufficient knowledge of event frequencies and severity parameters, as well as present and future boundary conditions. Expert judgment can be used to take account of these epistemic uncertainties which might nevertheless be difficult to limit, especially with respect to event or scenario probabilities (Beven et al., 2015).

References

ATC, 1978. Tentative provisions for the development of seismic design regulations of buildings. Report ATC 3-06, Applied Technology Council.

Baban, S.M.J., 2014. Managing Geo-Based Challenges: World-Wide Case Studies and Sustainable Local Solutions. Springer, Cham, p. 58.

Baer, E.M., 2007. Teaching quantitative concepts in floods and flooding. Science Education Resource Center at Carlton College (SERC), Northfield, MN.

Baker, J.W., 2008. An Introduction to Probabilistic Seismic Hazard Analysis (PSHA), White Paper, Version 1.3. Stanford University, California.

Beven, K.J., Aspinall, W.P., Bates, P.D., Borgomeo, E., Goda, K., Hall, J.W., Page, T., Phillips, J.C., Rougier, J.T., Simpson, M., Stephenson, D.B., Smith, P.J., Wagener, T., Watson, M., 2015. Epistemic uncertainties and natural hazard risk assessment—Part: 1 A review of the issues. Nat. Hazards Earth Syst. Sci. 3, 7333.

Bommer, J., 2002. Deterministic vs. probabilistic seismic hazard assessment: an exaggerated and obstructive dichotomy. J. Earthquake Eng. 6, 43.

Campedel, M., Cozzani, V., Garcia-Agreda, A., Salzano, E., 2008. Extending the quantitative assessment of industrial risks to earthquake effects. Risk Anal. 28, 1231.

CEN, 2004. Eurocode 8 Design of Structures for earthquake resistance, European Committee for Standardization, Brussels.

Cornell, C.A., 1968. Engineering seismic risk analysis. Bull. Seismol. Soc. Am. 58, 1583.

de Moel, H., van Alphen, J., Aerts, J.C.J.H., 2009. Flood maps in Europe—methods, availability and use. Nat. Hazards Earth Syst. Sci. 9, 289.

ESPON, 2006. The Spatial Effects and Management of Natural and Technological Hazards in Europe—ESPON 1.3.1, European Spatial Planning Observation Network. www.espon.eu

European Union, 2000. Directive 2000/60/EC of the European Parliament and of the Council of 23 October 2000 establishing a framework for Community action in the field of water policy. Off. J. Eur. Commun., L327/43.

European Union, 2007. Directive 2007/60/EC of the European Parliament and of the Council of 23 October 2007 on the assessment and management of flood risks. Off. J. Eur. Union, L288/27.

Geist, E.L., Lynett, J.P., 2014. Source processes for the probabilistic assessment of tsunami hazards. Oceanography 27, 86–93.

IAEA, 2003. Seismic Design & Qualification for Nuclear Power Plants - Safety Guide, Safety Standards Series No. NS-G-1.6, International Atomic Energy Agency, Vienna.

IPCC, 2007. In: Pachauri, R.K., Reisinger, A. (Eds.), "Climate Change 2007 Synthesis Report" Contribution of Working Groups I, II and III to the Fourth Assessment Report of the Intergovernmental Panel on Climate Change, Geneva, Switzerland.

ISO, 2000. ISO 15686-1 Buildings and constructed assets—service life planning—Part 1 General principles, International Organization for Standardization, Geneva.

Koshimura, S., Oie, T., Yanagisawa, H., Imamura, F., 2009. Developing fragility functions for Tsunami270 damage estimation using numerical model and post-tsunami data from Banda Aceh, Indonesia. Coastal Eng. J. 51, 243.

Kramer, S.L., 1996. Geotechnical Earthquake Engineeringvol. 80 Prentice Hall, Upper Saddle River, New Jersey.

Krausmann, E., Cozzani, V., Salzano, E., Renni, E., 2011. Industrial accidents triggered by natural hazards: an emerging risk issue. Nat. Hazards Earth Syst. Sci. 11, 921.

Krausmann, E., Cruz, A.M., Affeltranger, B., 2010. The impact of the 12 May 2008 Wenchuan earthquake on industrial facilities. J. Loss Prev. Process Ind. 23, 242.

Lanzano, G., Salzano, E., Santucci de Magistris, F., Fabbrocino, G., 2013. Seismic vulnerability of natural gas pipelines. Reliab. Eng. Syst. Saf. 117, 73.

Lanzano, G., Salzano, E., Santucci de Magistris, F., Fabbrocino, G., 2014a. Seismic vulnerability of gas and liquid buried pipelines. J. Loss Prev. Process Ind. 28, 72.

Lanzano, G., Santucci de Magistris, F., Fabbrocino, G., Salzano, E., 2015. Seismic damage to pipelines in the framework of Natech risk assessment. J. Loss Prev. Process Ind. 33, 159.

Lanzano, G., Santucci de Magistris, F., Salzano, E., Fabbrocino, G., 2014b. Vulnerability of industrial components to soil liquefaction. Chem. Eng. Trans. 36, 421.

Lorito, S., Selva, J., Basili, R., Romano, F., Tiberti, M.M., Piatanesi, A., 2015. Probabilistic hazard for seismically induced tsunamis: accuracy and feasibility of inundation maps. Geophys. J. Int. 2001, 574.

Luco, N., Ellingwood, B.R., Hamburger, R.O., Hooper, J.D., Kimball, J.K., Kircher, C.A., 2007. Risk-targeted versus current seismic design maps for the conterminous United States. In: Proceedings of the 2007 Convention of the Structural Engineers Association of California (SEAOC).

Mas, E., Koshimura, S., Suppasri, A., Matsuoka, M., Matsuyama, M., Yoshii, T., Jimenez, C., Yamazaki, F., Imamura, F., 2012. Developing Tsunami fragility curves using remote sensing and survey data of the 2010 Chilean Tsunami in Dichato. Nat. Hazards Earth Syst. Sci. 12, 2689.

NBCC, 2005. National Building Code of Canada. Institute for Research Construction, National Research Council of Canada, Ottawa, Canada.

NEHRP, 2003. NEHRP Recommended Provisions for Seismic Regulations for New Buildings and Other Structures—Part I Provisions, US Federal Emergency Management Authority, Washington, DC.

NEHRP, 2009. NEHRP Recommended Seismic Provisions for New Buildings and Other Structures. Part I—Provisions, US Federal Emergency Management Authority, Washington, DC.

Nelson, S.A., 2015. Flooding Hazards, Prediction & Human Intervention. Tulane University, Department of Earth and Environmental Sciences, New Orleans, USA. http://www.tulane.edu/~sanelson/Natural_Disasters/floodhaz.htm

Nelson, S.A., 2012. Tsunami. Tulane University, Department of Earth and Environmental Sciences, New Orleans, USA. http://www.tulane.edu/~sanelson/Natural_Disasters/tsunami.htm

NTC, 2008. Italian Building Code, DM 14, 2008.

OPCM, 2003. Ordinanza del Presidente del Consiglio dei Ministri n. 3274 del 20 marzo 2003, Gazzetta Ufficiale della Repubblica italiana n.105 dell'8 maggio 2003, Primi elementi in materia di criteri generali per la classificazione sismica del territorio nazionale e di normative tecniche per le costruzioni in zona sismica.

Panico, A., Basco, A., Lanzano, G., Pirozzi, F., Santucci de Magistris, F., Fabbrocino, G., Salzano, E., 2016. Evaluating the structural priorities for the seismic vulnerability of civilian and industrial wastewater treatment plants, Saf. Sci., in press.

Pellegrino, A.M., Scotto di Santolo, A., Evangelista, A., Coussot, P., 2010. Rheological behaviour of pyroclastic debris flow. Third International Conference on Monitoring, Simulation, Prevention and Remediation of Dense and Debris Flow "Debris Flow III," May 24–26, Milan, Italy.

Pellegrino, A.M., Scotto di Santolo, A., Schippa, L., 2015. An integrated procedure to evaluate rheological parameters to model debris flows. Eng. Geol. 196, 88.

PGEC, 2010. Methodology for probabilistic tsunami hazard analysis: trial application for the Diablo Canyon Power Plant Site. Pacific Gas & Electric Company, Berkeley, CA.

Rikitake, T., Aida, I., 1988. Tsunami hazard probability in Japan. Bull. Seismol. Soc. Am. 78, 1268.

Salzano, E., Basco, A., Busini, V., Cozzani, V., Renni, E., Rota, R., 2013. Public awareness promoting new or emerging risk: industrial accidents triggered by natural hazards. J. Risk Res. 16, 469.

Salzano, E., Garcia-Agreda, A., Di Carluccio, A., Fabbrocino, G., 2009. Risk assessment and early warning systems for industrial facilities in seismic zones. Reliab. Eng. Syst. Saf. 94, 1577.

Salzano, E., Iervolino, I., Fabbrocino, G., 2003. Seismic risk of atmospheric storage tanks in the framework of quantitative risk analysis. J. Loss Prev. Process Ind. 16, 403.

Suppasri, A., Mas, E., Charvet, I., Gunasekera, R., Imai, K., Fukutani, Y., Abe, Y., Imamura, F., 2013. Building damage characteristics based on surveyed data and fragility curves of the 2011 Great East Japan Tsunami. Nat. Hazards 66, 319.

Thio, H.K., Somerville, P., Polet, J., 2010. Probabilistic Tsunami Hazard in California. Pacific Earthquake Engineering Research Center, PEER Report 2010/108. University of California, Berkeley, CA.

Technological Hazard Characterization

V. Cozzani, E. Salzano
Department of Civil, Chemical, Environmental, and Materials Engineering,
University of Bologna, Bologna, Italy

In the framework of Natech risk assessment, technological hazard refers to a potential for the occurrence of adverse effects due to the release of hazardous substances, from process or storage equipment, caused by natural-hazard impacts. It is important to identify the equipment that acts as a hazard source, and to rank the hazard in terms of the type and amount of hazardous substances it contains, its operating conditions, as well as its vulnerability with respect to natural hazards. This chapter examines these aspects to support the prioritization of specific measures for Natech accident prevention and mitigation, and to focus the application of detailed quantitative risk-analysis techniques to a limited set of accident scenarios or equipment items that are considered to be the most critical.

6.1 INTRODUCTION

The United Nations Office for Disaster Risk Reduction defines a technological hazard as (UNISDR, 2015):

> A hazard originating from technological or industrial conditions, including accidents, dangerous procedures, infrastructure failures or specific human activities, that may cause loss of life, injury, illness or other health impacts, property damage, loss of livelihoods and services, social and economic disruption, or environmental damage.

The comment to this definition explicitly mentions the Natech hazards that originate from the impacts of natural events on hazardous industrial installations, such as nuclear power plants, or chemical processing, and storage facilities. As was already discussed in the previous chapters, the release of hazardous substances due to process upsets or structural failure triggered by natural hazards can affect plant workers, the population, property, and the environment both in- and outside the area impacted by the natural event.

In the framework of Natech risk assessment, the evaluation of technological hazards serves the purpose of identifying the potential sources of adverse effects involving the release of hazardous substances, following the impact of one or multiple natural events. This starts with the assessment of the hazard that an equipment item containing a hazardous substance poses, regardless of the accident trigger. However, when dealing with natural hazards, several limitations present themselves. In particular, the structural capacity, i.e., the absolute strength of the system beyond the expected or design loads or, in other words, the structural vulnerability of the system with respect to natural events is crucial and needs to be considered in Natech risk assessment.

With respect to Natech risk, the potential technological hazard sources should be prioritized by ranking process and storage equipment on the basis of three main factors:

1. The hazard of the substance or mixture of substances and the amount of the substance or mixture of substances (stored, produced, or transported) present in the unit or equipment item considered.
2. The physical state of the substance contained in the equipment.
3. The structural vulnerability of the vessel (equipment item) with respect to the natural event.

These three factors will be discussed in the following sections with specific reference to natural-hazard impacts.

6.2 SUBSTANCE HAZARD

The hazard associated with a substance is strictly related to the intrinsic capacity of the substance, or mixture of substances, to produce harm to people, property, and the environment. This hazard is clearly general and is not specific to Natech risks. Hence, the substance information in the European REACH Regulation (European Union, 2006), which regulates the registration, evaluation, authorization, and restriction of chemicals in the European Union, can be used for the evaluation of the technological hazard.

This information can be combined with hazard categories from the European Regulation on classification, labeling, and packaging (CLP) of substances and mixtures (ECHA, 2015; European Union, 2008), and substance data and related threshold quantities in the Seveso Directive on the control of major-accident hazards involving dangerous substances (European Union, 1997) and its amendments (European Union, 2003, 2012). The Seveso Directive also includes a list of particularly problematic substances that are named explicitly and for which separate qualifying quantities apply. Table 6.1 shows the list of dangerous-substance categories and associated thresholds for lower- and upper-tier Seveso establishments as per the Seveso Directive. Further details can be found in the cited Directives and Regulations. Eventually, the substances to be considered in the hazard assessment are only those that are stored or processed in sufficient amounts to produce damage.

Analyses of past Natech accidents showed that for Natech risk assessment special attention needs to be paid to atypical scenarios caused by the interaction between

Table 6.1 List of Substance-Hazard Categories According to the CLP Regulation and Threshold Amounts for Lower- and Upper-Tier Installations as per the Seveso Directive (European Union, 2008, 2012)

Hazard Categories	Lower-Tier	Upper-Tier
	Amount (*t*)	
Section "H"—HEALTH HAZARDS		
H1 ACUTE TOXIC Category 1, all exposure routes	5	20
H2 ACUTE TOXIC, Category 2, all exposure routes; Category 3, inhalation exposure route	50	200
H3 STOT SPECIFIC TARGET ORGAN TOXICITY—SINGLE EXPOSURE STOT SE Category 1	50	200
Section "P"—PHYSICAL HAZARDS		
P1a EXPLOSIVES, Unstable explosives or Explosives, Division 1.1, 1.2, 1.3, 1.5, or 1.6, or substances or mixtures having explosive properties according to method A.14 of Regulation (EC) No 440/2008 and do not belong to the hazard classes organic peroxides or self-reactive substances and mixtures	10	50
P1b EXPLOSIVES, Explosives, Division 1.4	50	200
P2 FLAMMABLE GASES Flammable gases, Category 1 or 2	10	50
P3a FLAMMABLE AEROSOLS, "Flammable" aerosols Category 1 or 2, containing flammable gases Category 1 or 2 or flammable liquids Category 1	150 (net)	500 (net)
P3b FLAMMABLE AEROSOLS, "Flammable" aerosols Category 1 or 2, not containing flammable gases, Category 1 or 2 nor flammable liquids Category 1)	5,000 (net)	50,000 (net)
P4 OXIDISING GASES Oxidizing gases, Category 1	50	200
P5a FLAMMABLE LIQUIDS, Flammable liquids, Category 1, or Flammable liquids Category 2 or 3 maintained at a temperature above their boiling point, or other liquids with a flash point ≤60°C, maintained at a temperature above their boiling point	10	50
P5b FLAMMABLE LIQUIDS, Flammable liquids Category 2 or 3 where particular processing conditions, such as high pressure or high temperature, may create major-accident hazards, or other liquids with a flash point ≤60°C where particular processing conditions, such as high pressure or high temperature, may create major-accident hazards	50	200
P5c FLAMMABLE LIQUIDS, Flammable liquids, Categories 2 or 3 not covered by P5a and P5b	5,000	50,000
P6a SELF-REACTIVE SUBSTANCES AND MIXTURES and ORGANIC PEROXIDES, self-reactive substances and mixtures, Type A or B or organic peroxides, Type A or B	10	50
P6b SELF-REACTIVE SUBSTANCES AND MIXTURES and ORGANIC PEROXIDES, self-reactive substances and mixtures, Type C, D, E, or F or organic peroxides, Type C, D, E, or F	50	200

(Continued)

Table 6.1 List of Substance-Hazard Categories According to the CLP Regulation and Threshold Amounts for Lower- and Upper-Tier Installations as per the Seveso Directive (European Union, 2008, 2012) (cont.)

	Lower-Tier	Upper-Tier
Hazard Categories	Amount (t)	
P7 PYROPHORIC LIQUIDS AND SOLIDS Pyrophoric liquids, Category 1 Pyrophoric solids, Category 1	50	200
P8 OXIDISING LIQUIDS AND SOLIDS Oxidizing Liquids, Category 1, 2, or 3, or oxidizing solids, Category 1, 2, or 3	50	200
Section "E"—ENVIRONMENTAL HAZARDS		
E1 Hazardous to the Aquatic Environment in Category Acute 1 or Chronic 1	100	200
E2 Hazardous to the Aquatic Environment in Category Chronic 2	200	500
Section "O"—OTHER HAZARDS		
O1 Substances or mixtures with hazard statement EUH014	100	500
O2 Substances and mixtures which in contact with water emit flammable gases, Category 1	100	500
O3 Substances or mixtures with hazard statement EUH029	50	200

the substance hazard and the specific features of the natural event causing the loss of containment. In particular, accident scenarios involving the release of hazardous substances into water are quite frequent for some Natech events (floods, heavy rain, tsunami, etc.). In this context, hazards related to water contamination but also to substance reactivity with water may play an important role in influencing and possibly aggravating the risks due to the technological hazard (cf. Chapter 3).

6.3 PHYSICAL STATE OF THE RELEASED SUBSTANCE

The physical state of the release plays an important role in establishing the hazard due to Natech events. Several physical states are of relevance in determining the substance hazard: liquefied gases under pressure, cryogenic liquefied gases, compressed gases, liquids, and solids.

Liquefied gases under pressure are fluids that are in the liquid phase at a temperature that is higher than their boiling point at atmospheric pressure. Two different options exist: (1) liquefied gases under pressure at ambient temperature (i.e., fluids that at ambient temperature and atmospheric pressure would be in the gas phase), and (2) fluids in the liquid phase at a temperature higher than ambient temperature and higher than their boiling point at atmospheric pressure.

Option (1) is a widely used storage strategy for many categories of chemicals and fuel, such as liquefied petroleum gas (LPG), chlorine, and ammonia. Option (2) is usually applied in certain processing activities of the chemical industry. Equipment containing liquids under pressure at high temperature is specific to process operations and not of bulk storage, it is less frequently used and has a lower inventory. Nevertheless, for both options, any failure of the equipment that contains hazardous materials may result in a rapid or almost instantaneous release of the substance to the atmosphere, thus producing severe accident scenarios due to the simultaneous release of vapor, a liquid aerosol, and, depending on the release conditions, the formation of a pool of boiling liquid.

If gas is liquefied by cooling the fluid below ambient temperature, a cryogenic liquid is obtained. *Cryogenic liquids* are liquefied gases that are kept in their liquid state at very low temperatures (usually at their boiling point at atmospheric pressure). All cryogenic liquids are gases at ambient temperature and atmospheric pressure. The most important example of this type of substance is liquefied natural gas (LNG). The main technological hazard is related to loss of containment of the substance. If a cryogenic liquid is released to the environment, a sudden vaporization takes place until a pool of boiling liquid is formed. A vapor cloud originates from the initial vaporization and from the boiling pool. However, the energy needed for the vaporization is supplied from the environment (ground or air). This usually decreases the rate of evaporation, thus affecting the dispersion mode. However, when marine structures are of concern (as for LNG, e.g., LNG terminals or jetties) the rapid heat exchange with the sea or river/lake water may affect the evaporation and trigger anomalous accidents (Bubbico and Salzano, 2009).

Compressed gas releases will immediately generate a gas cloud. However, due to the limited inventory of storage systems, caused by the low density of compressed gases, the actual relevance and overall severity of such releases needs to be specifically assessed.

When *liquids* are released, hazards are created due to the entrainment of vapors in the air that comes in contact with the liquid. Liquid evaporation takes place, and a cloud of vapors may be formed. Nevertheless, the intensity of evaporation and the concentration of vapors in the cloud are expected to be much lower than for liquefied gases, due to the difference in driving force available for evaporation.

The hazard related to substances in the *solid* state is essentially associated with the chemical properties of the substance. However, dust explosions and fires are mainly linked with the physical characteristics. Hence, a "fine dust" physical state should be considered.

The detailed analysis of release scenarios is out of the scope of this chapter. The reader is referred to comprehensive publications on the topic to obtain more detailed information on the qualitative and quantitative characterization of events involving the loss of containment of chemical substances in different physical states (Mannan, 2005; van Den Bosch and Weterings, 2005).

In the case of Natech scenarios, the possibility that liquid pools or released solid substances come in contact with and are displaced by water needs to be taken into account in the risk analysis. This may cause specific accident scenarios that are usually disregarded when loss of containment takes place due to internal failures of the system.

6.4 EQUIPMENT VULNERABILITY

The vulnerability of an equipment item in a Natech event derives from its structural features. As such, its vulnerability can be obtained from detailed structural modeling only. Nevertheless, in the framework of quantitative risk analysis, the introduction of a simplified approach to the assessment of equipment vulnerability with respect to natural-hazard loading is useful. Ranking methodologies of process units are thus needed to help prioritize intervention, design prevention and mitigation systems and measures, be they technical or organizational, and restrict the application of detailed analysis techniques (QRA) to a limited number of critical equipment items.

On the path to defining a ranking procedure and the propensity to cause an accident in a given technological system, it is worth considering that any process which converts raw materials or intermediate products to final products, or any transportation system of fluids, should first be ranked by virtue of the specific hazard of the substance involved. Simply put, a large storage tank containing water cannot be considered as hazardous as a similar tank containing flammable materials, whatever its scale or construction characteristics. Besides, for the given substance, several possible equipment types might exist, each of which with different structural characteristics, functions, and scale.

Various approaches for classifying equipment using different categories are available. Based on several past analyses, equipment was categorized in three classes with respect to the design standard: (1) atmospheric equipment (storage tank and process units), (2) pressurized equipment (cylindrical buried, cylindrical aboveground, spheres), and (3) pipeline systems. The following sections will be devoted to these three types of equipment only.

6.4.1 Atmospheric Equipment

Atmospheric equipment includes storage tanks and process units adopted for a range of applications, such as distillation, separation, extraction, etc. Due to their capacity and wide diffusion, atmospheric storage tanks are the most relevant type of equipment. They are constructed worldwide following the American Petroleum Institute standard API 650 (American Petroleum Institute, 2007) and are typically vertical cylinders. Other atmospheric process equipment, such as distillation towers, or cyclones, are also designed to similar codes and standards, they have, however, slender geometry.

From a structural point of view, all these equipment types are generally built by using carbon steel or stainless steel, with a typical maximum allowable working pressure (MAWP) and corresponding failure pressure of few millibars. Shell thicknesses range from 5 mm to about 1 cm for some sections of jumbo tanks. Interestingly, studies have shown that earthquake-triggered structural damage involving water tanks is very similar to tanks containing hazardous materials and their behavior can be described using a similar methodology. Distillation and absorption columns, and similar tower units typically have an internal diameter greater than 0.1 m.

6.4.2 Pressurized Equipment

Pressurized equipment is often used for very hazardous substances and is typically cylindrical (buried or above-ground) or spherical. Their design and certification is governed worldwide by well-known design codes, such as the ASME Boiler and Pressure Vessel Code (ASME, 2015) in North America, and the EU Directive 2009/105/EC (European Union, 2009).

The most common hazardous system in the process industry is related to the storage of large amounts of liquefied gases, such as LPG (propane, butane, and their mixture), ethylene, or hydrogen. The shell thickness and the corresponding design pressure is clearly larger than for atmospheric equipment and may reach several centimeters for small equipment like chemical reactors. Pressure vessels are typically welded, and are not intended to be fired and subjected to an internal gauge pressure greater than 0.5 bars. The MAWP is generally lower than 30 bar, and the minimum and maximum working temperatures are, respectively, −50 and 300°C for steel or 100°C for aluminum or aluminum alloy vessels.

When large amounts of hazardous materials are stored in pressurized equipment, the vessels are often mounded or buried, to avoid any interaction with external effects in case of fire, explosion, or simple collisions with objects. In this case, several accident scenarios can be considered as not credible (and the technological hazard is therefore reduced to negligible values) due to the physical impossibility of accidents. Nevertheless, this kind of equipment may have been designed with external auxiliary systems and pipes, which should also be considered as a source of technological hazard.

Pressurized equipment typically also includes shell and tube heat exchangers, seal-less pumps with a specified maximum flow-rate greater than 0.5 m³/h, or reactors, and some elongated vessels (distillation).

6.4.3 Pipelines

Pipeline systems within industrial installations may be above-ground or buried. The pipe body can be continuous or segmented and is typically built from carbon or stainless steel when transporting hazardous substances.

For the evaluation of the technological hazard, the pipeline system has to be separated into transportation and distribution networks. The transportation network is generally used to transfer the liquid or gas from the production place to the industrial plants or urban distribution system.

With respect to gas, overland transportation pipelines generally operate at high pressure (>70 bars), in order to transfer a large amount of fluid per unit time. In the

United States, for example, the large-scale natural-gas transmission system includes around 300,000 km of high-strength, steel pipelines, with diameters between 0.6 and 0.9 m and pressures between 34 and 97 bars. With respect to the seismic vulnerability of pipeline systems, two large categories based on pipe diameter exist: (1) $D \geq 400$ mm for high-pressure transmission systems; and (2) $D < 400$ mm for distribution and low-pressure transmission systems (Lanzano et al., 2013).

For distribution systems the most common pipe materials are cast iron, ductile iron, steel, and polymers. Cast iron was largely used in the last century. This material shows high fragility and lacks ductility, which raises safety concerns. For these reasons, pipelines are nowadays made of ductile iron, steel, and plastic materials like polyvinylchloride, polyethylene (HDPE), and glass-reinforced fiber polymers. Other construction materials, such as concrete, are used for water and wastewater pipelines.

The damage patterns occurring in these structures are largely dependent on the material base properties and the joint detailing. For this reason, all the possible combinations of material and joints are typically divided into two categories: (1) continuous pipelines (CP) and (2) segmented pipelines (SP), or equivalently in brittle (SP) and ductile (CP) in terms of prefailure deformations.

Table 6.2 shows the main structural features, which are essential for gas and liquid pipelines. It is worth noting that hazardous materials (toxic, flammable) must be transported only in continuous pipelines, which have high strength and can tolerate large deformations before breaking and subsequent fluid release.

The choice of the joints is also a crucial issue in the seismic design of pipelines, particularly for those used for gas. In order to avoid that the pipeline joints perform as weak points, they must be designed aiming at restoring the continuity of the pipeline body in terms of strength and stiffness. This is achieved by mostly using welded joints, however, in some cases, mechanical and special joints are also used.

Among the different welding techniques, three are remarkable for steel pipelines: (1) oxyacetylene welding (OAW), (2) submerged arc welding (SAW), and (3) high-quality welding. In the past, the preferred welding type belonged to the first and second categories. In fact, the SAW gives a good strength recovery compared to OAW, which suffered extensive damage in earthquakes.

Table 6.2 Structural Features of Pipelines (Lanzano et al., 2014)

Pipelines	Materials	Joints	Damage Patterns
Continuous (CP)	Steel; polyethylene; polyvinylchloride; glass fiber reinforced polymer	Butt welded; welded slip; chemical weld; mechanical joints; special joints	Tension cracks; local buckling; beam buckling
Segmented (SP)	Asbestos cement; reinforced concrete; polyvinylchloride (PVC); vitrified clay; cast iron	Caulked joints; bell end spigot joints	Axial pull-out; crushing of bell end; crushing of spigot joints; circumferential failure; flexural failure

An important distinguishing factor for the hazard from pipeline systems is whether the pipeline is installed above or below ground (Lanzano et al., 2014). Generally, the burial depth of gas pipelines is in the range of 1–2 m. Pipelines with very large diameters might be buried deeper. For the above-ground case, the use of support structures is common. Gas pipes are frequently placed below ground level. The burying process is beneficial for two reasons. On the one hand, the surrounding soil protects the pipeline from above-ground hazards, such as natural events or accidents. Secondly, the lateral confinement provided by the soil, which increases with depth, reduces the likelihood of domino effects due to fire or explosion.

A simple measurement for pipeline performance and the associated technological hazard is the pipeline repair rate, RR (ALA, 2001).

6.4.4 Hazard Classification Based on Structural Features and Hazard of the Secondary Scenario

The analysis of the expected damage and the criticality of the associated accident scenarios can provide further indications on whether equipment is critical. In the framework of Natech risk analysis, the structural damage itself may have only a limited relevance. The severity of the accident following equipment damage and loss of containment is the element that should take priority in the analysis.

A first classification of equipment criticality may be derived from Table 6.3 for units containing substances with the hazard characteristics and nonnegligible substance amounts as shown in Table 6.1. Substance characteristics were combined with the expected structural vulnerability of the equipment and the relevance of the hazard related to its physical state.

The highest hazard is assumed for above-ground, pressurized equipment that contains liquefied gases. Any structural failure or loss of control of the associated industrial process, either related to anthropogenic causes or natural events, may result in

Table 6.3 Technology Hazard Matrix With 1 = Low Hazard and 5 = High Hazard

Equipment	Liquefied Gas	Compressed Gas	Cryogenic Liquid	Liquid	Fine Dusts
Pressurized (above-ground)	5	4	4	2	1
Pressurized (underground)	2	3	2	2	1
Atmospheric	—	—	5	3	3
Pipeline (above-ground)	4	3	4	2	1
Pipeline (underground)	3	2	3	1	—

Underground equipment is considered buried or mounded.

a large-scale accident due to the rapid release of content and the possible escalation to fire, explosion, toxic dispersion, or even to environmental disaster. The score is higher than for nonliquefied gases because of the large difference in density, and hence mass per volume. A similar reasoning can be used to explain the high hazard score for above-ground pipelines, which contain, however, lower amounts of hazardous substances. Cryogenic liquids contained in atmospheric vessels also result in the maximum hazard score due to the vessels' intrinsic fragility in case of internal pressurization.

When buried or mounded, both liquefied and cryogenic liquids are less hazardous due to the intrinsic protection afforded by the mound or surrounding soil. Compressed gas may be slightly more hazardous if released through thin layers of soil or structures.

Where liquid substances are a concern, the hazard is related to their flammability and toxicity, and also to environmental issues. Atmospheric equipment is typically utilized for high-capacity storage of hazardous liquids, and this leads in turn to higher hazards. Although having a lower hazard score for liquids, pressurized equipment is typically used for very hazardous materials. Consequently, the failure and following loss of containment can also result in nonnegligible hazards.

Finally, the highest hazard score for fine dusts is associated with large-scale atmospheric equipment, such as silos or mills. In this case, fire or explosions may be induced by external causes which can result in severe accident scenarios.

A more detailed evaluation of the technological hazard should, however, also take into account the intensity of the loss of containment following equipment damage and the specific hazard of the material released. A useful approach to assess escalation thresholds is the description of secondary target damage by a discrete number of structural Damage States (DSs) and of Loss Intensities (LIs) following the scheme originally introduced to obtain a cost estimate of damage caused by explosions (Tam and Corr, 2000) or by natural events (HAZUS, 1997). The structural DS of equipment may, for instance, be described by two classes: DS1, equivalent to minor damage to the structure or to auxiliary equipment, and DS2, with intense damage or even total collapse of the structure.

The shift to DSs due to natural-hazard impact may also be associated to a loss of containment, whose intensity is among the most important factors affecting the relevance of the Natech scenario. In fact, increasing LIs usually result in an increase of the severity of the loss-of-containment scenario and in a decrease of the time available for successful mitigation. The LIs following vessel damage may then be represented by a discrete number of LI categories or so-called risk states RS. Following the approach used in the *Purple Book* (Uijt de Haag and Ale, 1999), three LI categories were defined: (1) LI1: "minor loss," defined as the partial loss of inventory, or the total loss of inventory in a time interval higher than 10 min from the impact of the blast wave; (2) LI2: "intense loss," defined as the total loss of inventory in a time interval between 1 and 10 min; and (3) LI3: "catastrophic loss," defined as the "instantaneous" complete loss of inventory (complete loss in a time interval of less than 1 min).

As a first approximation, it is obvious that LI1 losses are usually associated to DS1, whereas loss states LI2 and LI3 can in general be associated to a DS2 state. However, a further factor that should be taken into account is the hazard posed by the substance released from the damaged equipment item. In particular, if the same LI is considered, in case of volatile releases, toxic substances may cause more severe accident scenarios than flammable substances. On the other hand, for nonvolatile releases, flammable substances represent more severe hazards than toxic substances. Table 6.4 shows the expected Natech scenarios and the associated severity for

Table 6.4 Expected Secondary Scenarios and Estimated Severity for Different Target Equipment and Loss Intensity Classes[a]

	Expected Secondary Events for Different Target Equipment			
Loss Intensity	Atmospheric Equipment	Pressurized Equipment	Elongated Equipment	Auxiliary Equipment
LI1—Flammable	Minor pool fire	Minor jet fire	Minor pool fire; minor flash fire	Minor pool fire; minor flash fire
LI1—Toxic	Minor evaporating pool	Boiling pool; jet toxic dispersion	Minor boiling pool; toxic dispersion	Minor evaporating pool
LI2—Flammable	Pool fire; flash fire; VCE	Jet fire; flash fire; VCE	Pool fire; flash fire; VCE	Minor pool fire; minor flash fire
LI2—Toxic	Evaporating pool; toxic dispersion	Boiling pool; jet toxic dispersion	Boiling pool; toxic dispersion	Minor evaporating pool
LI3—Flammable	Pool fire; flash fire; VCE	BLEVE/fireball; flash fire; VCE	Pool fire; flash fire; VCE	Minor pool fire; minor flash fire
LI3—Toxic	Evaporating pool; toxic dispersion	Boiling pool; jet toxic dispersion	Boiling pool; toxic dispersion	Evaporating pool; minor toxic dispersion
	Expected Severity			
Loss Intensity	Atmospheric Equipment	Pressurized Equipment	Elongated Equipment	Auxiliary Equipment
LI1—Flammable	Low	High	Low	Low
LI1—Toxic	Low	High	High	Low
LI2—Flammable	High	High	High	Low
LI2—Toxic	High	High	High	Low
LI3—Flammable	High	High	High	Low
LI3—Toxic	High	High	High	High

[a]VCE, Vapor cloud explosion; BLEVE, boiling liquid expanding vapor explosion. "Flammable" and "Toxic" refer to the substance in the secondary vessel damaged by the blast wave.
Adapted from Cozzani et al. (2006).

different LIs and DSs, taking into account the hazard posed by the released substance. This concept of DSs and associated risk states is further explored in Chapter 7.

6.5 CONCLUSIONS

When determining the technological hazard, the danger associated with the intrinsic chemical and physical properties related to the processed and stored substances cannot be neglected. Equipment that contains flammable/toxic, highly flammable/toxic, or extremely flammable/toxic substances according to the CLP Regulation should certainly be considered as relevant source of potentially severe accidents. In addition, a substance's physical state (gas, liquid, solid) and the equipment's operating conditions, which depend on the specific processing or storage activity, are also of extreme importance. In this context, hazard matrices can be a useful tool for ranking the hazard related to process equipment and to identify the units that have to be taken into account in the Natech risk analysis.

References

ALA, 2001. Seismic Fragility Formulations for Water Systems, American Lifelines Alliance. American Society of Civil Engineers and US Federal Emergency Management Agency, Washington, DC.

American Petroleum Institute, 2007. API 650—Welded Steel Tanks for Oil Storage, 11th ed. American Petroleum Institute, Washington, DC.

ASME, 2015. Boiler and Pressure Vessel Code—An International Code, 2015 Ed. The American Society of Mechanical Engineers, New York.

Bubbico, R., Salzano, E., 2009. Acoustic analysis of blast waves produced by rapid phase transition of LNG. Saf. Sci. 47 (4), 511.

Cozzani, V., Gubinelli, G., Salzano, E., 2006. Escalation thresholds in the assessment of domino accidental events. J. Hazard. Mater. 129, 1.

ECHA, 2015. Guidance on the Application of the CLP Criteria—Guidance to Regulation (EC) No 1272/2008 on Classification, Labelling and Packaging (CLP) of Substances and Mixtures, Version 4.1. European Chemicals Agency, Helsinki.

European Union, 1997. Council Directive 96/82/EC of 9 December 1996 on the control of major-accident hazards involving dangerous substances, Off. J. Eur. Commun., L 10/13.

European Union, 2003. Directive 2003/105/EC of the European Parliament and of the Council of 16 December 2003 Amending Council Directive 96/82/EC, Off. J. Eur. Union, L 345/97.

European Union, 2006. Regulation (EC) No 1907/2006 of the European Parliament and of the Council of 18 December 2006 Concerning the Registration, Evaluation, Authorization and Restriction of Chemicals (REACH), establishing a European Chemicals Agency, amending Directive 1999/45/EC and repealing Council Regulation (EEC) No. 793/93 and Commission Regulation (EC) No. 1488/94 as well as Council Directive 76/769/EEC and Commission Directives 91/155/EEC, 93/67/EEC, 93/105/EC and 2000/21/EC., Off. J. Eur. Union, L 396/1.

European Union, 2008. Regulation (EC) No 1272/2008 of the European Parliament and of the Council of 16 December 2008 on classification, labelling and packaging of substances and mixtures, amending and repealing Directives 67/548/EEC and 1999/45/EC, and amending Regulation (EC) No 1907/2006, Off. J. Eur. Union, L 353/1.

European Union, 2009. Directive 2009/105/EC of the European Parliament and of the Council of 16 September 2009 relating to simple pressure vessels, Off. J. Eur. Union, L 264/12.

European Union, 2012. Directive 2012/18/EU of the European Parliament and of the Council of 4 July 2012 on the control of major-accident hazards involving dangerous substances, amending and subsequently repealing Council Directive 96/82/EC, Off. J. Eur. Union, L 197/1.

HAZUS, 1997. Earthquake Loss Estimation Methodology, Technical Manual, National Institute of Building Science, Risk Management Solutions. HAZUS, Menlo Park, CA.

Lanzano, G., Salzano, E., Santucci de Magistris, F., Fabbrocino, G., 2013. Seismic vulnerability of natural gas pipelines. Reliab. Eng. Syst. Saf. 117, 73.

Lanzano, G., Salzano, E., Santucci de Magistris, F., Fabbrocino, G., 2014. Seismic vulnerability of gas and liquid buried pipelines. Reliab. Eng. Syst. Saf. 28, 72.

Mannan, S., 2005. Lees' Loss Prevention in the Process Industries, third ed. Elsevier, Oxford.

Tam, V.H.Y., Corr, B., 2000. Development of a limit state approach for design against gas explosions. J. Loss Prev. Process Ind. 13, 443.

Uijt de Haag, P.A.M., Ale, B.J.M., 1999. Guidelines for Quantitative Risk Assessment. Purple Book. Committee for the Prevention of Disasters, The Hague.

UNISDR, 2015. Terminology on Disaster Risk Reduction. United Nations Office for Disaster Risk Reduction, Geneva, https://www.unisdr.org/we/inform/terminology.

van Den Bosch, C.J.H., Weterings, R.A.P.M., 2005. Methods for the calculation of physical effects. Yellow Book. third ed Committee for the Prevention of Disasters, The Hague.

Natech Risk and Its Assessment

E. Krausmann
European Commission, Joint Research Centre, Ispra, Italy

Risk analysis is a powerful tool for estimating the risk level originating from a hazardous industrial activity. This chapter briefly introduces the industrial risk–assessment process and proposes a methodology for Natech risk assessment based on the conventional QRA procedure, illustrating the individual steps of the process. It also discusses the accompanying data requirements and common applications of the results.

7.1 GENERAL CONSIDERATIONS

In most countries a risk assessment is required to identify and minimize public-health threats or potential environmental effects of a proposed hazardous industrial activity, both during normal operation and in accident situations. Identification of potentially significant effects may lead to a modification of the proposed action to reduce risks or to the consideration of alternatives. The potential impacts of natural hazards are generally taken into account to some extent in the design and construction of facilities by adhering to dedicated codes and standards, although the impacts are not usually fully considered in risk-management plans.

However, as already discussed in Section 4.2.1, there are questions related to the adequacy of the design basis of hazardous installations against natural-hazard loading. The primary goal of natural-hazard resilient design for buildings and other structures (e.g., against earthquakes or high winds) is to prevent building collapse and therefore to guarantee life safety. In order to reduce Natech risk the preservation of the structural integrity is not sufficient; the avoidance of loss of containment (LOC) of hazardous materials must also be considered. In addition, it is usually unclear which level of damage or failure (possibly including LOC) is to be expected above the design-basis loading (Krausmann, 2016).

Consequently, the adoption of performance-based design against natural-hazard impacts is recommended for critical structures and buildings within an industrial plant. Performance-based standards require that safety-relevant buildings, equipment, and systems satisfy performance criteria (e.g., control rooms remain operational after a design-basis natural event) with respect to materials, equipment, and design and construction methods (Cruz and Okada, 2008).

Natech Risk Assessment and Management. http://dx.doi.org/10.1016/B978-0-12-803807-9.00007-3

Risk analysis is the tool of choice for estimating the level of risk produced by a hazardous activity. Quantitative risk analysis in particular allows the identification of system weaknesses, the prioritization of safety measures in terms of their yield for risk reduction, or the estimation of a facility's overall risk level, summarized in a risk figure. This risk figure can then be compared with prescribed target levels, where existing, to show that risks are adequately controlled in fulfillment of regulatory requirements.

7.2 THE INDUSTRIAL RISK–ASSESSMENT PROCESS

The qualitative or quantitative assessment of the risk from an industrial operation is an important step in controlling this risk. The assessment process comprises five steps: (1) the identification of the hazard(s) and failure mechanisms; (2) the estimation of the failure frequencies; (3) consequence analysis to understand the impact of failure in terms of overpressure, heat radiation, and dispersion of toxic materials; (4) risk integration or recomposition where likelihood and consequence information is combined to express total risk; and (5) risk evaluation where the risk estimate is compared to tolerability or acceptability targets or criteria (Cox, 1998). The first four steps are also referred to as risk analysis.

Qualitative risk analysis is often used as a first step in industrial risk assessment as it requires relatively little effort and no specific expertise in risk analysis. The risk is estimated by defining severity categories for both the event frequencies and the consequences which are combined in a so-called risk matrix (Table 7.1). This is a simple but effective method for obtaining an overview of which hazards need to be prioritized to reach predefined risk-reduction targets.

Quantitative risk analysis or QRA is a powerful technique for estimating industrial risk, but its application is complex and time consuming. A well-accepted framework for QRA exists, as well as specific tools for its practical implementation (e.g., Uijt de Haag and Ale, 1999; Delvosalle et al., 2004). QRA aims at assigning numbers to the likelihood and the consequences of a failure case, while at the same time considering the totality of all possible events (Cox, 1998). The risk estimate is then commonly represented as individual risk or societal risk (F–N curves), examples of which are shown in Figs. 7.1 and 7.2.

Individual risk is the probability for an individual to suffer ill effects (e.g., death or injury) at a specific point around a hazardous facility per given time period (Christou, 1998a). Connecting the points for which the individual risk has the same value yields so-called isorisk curves or individual risk contours. These curves are independent of the population density around the hazardous installation. Where a group of people is at risk from an industrial activity the risk indicator of choice is societal risk. F–N curves plot the cumulative frequency (F) of different accident scenarios against the number (N) of potential casualties— usually fatalities—associated with these scenarios. It is important to note that in F–N curves N represents the number of casualties that could be equalled or exceeded. Since the calculation of F–N curves

Table 7.1 Example of a Risk Matrix

Frequency	Consequence				
	Very low	Low	Moderate	High	Very high
Likely (> 0.1 yr^{-1})					
Possible (10^{-1}–10^{-2} yr^{-1})					
Unlikely (10^{-2}–10^{-3} yr^{-1})					
Very unlikely (10^{-3}–10^{-4} yr^{-1})					
Remote (< 10^{-4} yr^{-1})					

Red denotes areas of intolerable risk, orange and yellow indicate areas where risk reduction measures should be implemented, and green shows areas where risk is broadly acceptable.
Adapted from Cox (1998).

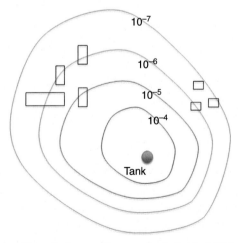

FIGURE 7.1 Example of Isorisk Curves Showing the Individual Risk Around a Tank Containing a Hazardous Substance

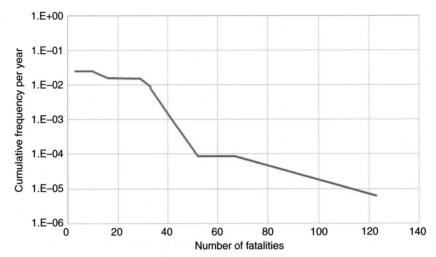

FIGURE 7.2 **Example of a Societal Risk Curve**

requires consideration of all potential accident scenarios, it is a resource-intensive exercise (HSE, 2009). The Center for Chemical Process Safety of the American Institute of Chemical Engineers has issued detailed guidelines on quantitative risk analysis for the chemical process industry (CCPS, 2000).

Like other methods for risk analysis, QRA is subject to a number of limitations, many of which arise from uncertainty in the input data or lack thereof, the used models, and the quality of the analysis. A lack of skilled human resources and time can also negatively impact the QRA. These limitations must be understood and considered when using QRA for decision-making purposes.

Regardless of the approach chosen, extensions to both qualitative and quantitative risk analyses need to be made to take into account the characteristics of Natech events. Time constraints and data availability permitting, the quantitative methodology is preferable to a qualitative approach. Therefore, in the following sections we will discuss a quantitative methodology for Natech risk analysis.

7.3 THE NATECH RISK–ASSESSMENT PROCESS

The conventional QRA process lends itself to Natech risk assessment. This requires the extension of the QRA methodology to include specific equipment damage models and consideration of the possibility of simultaneous loss-of-containment events at several units, an important characteristic of Natech accidents in general. Simple damage models are available for a limited number of equipment categories (storage tanks, some types of process equipment) and in particular for earthquake impact. Inclusion of these damage models in QRA case studies clearly showed the importance of considering earthquake-triggered accident scenarios for ensuring the safety of the facility itself and the surrounding population and environment (Antonioni et al., 2007). Therefore, natural hazards can be important risk contributors at hazardous facilities and must be adequately considered in the risk-analysis process.

7.3.1 Input

Natech risk assessment requires a significant amount of input information to evaluate the interaction of the natural hazard with the industrial system, as well as its possible consequences. The following list provides an overview of the data specifically required for seismic Natech risk analysis. It is, however, equally applicable to Natech risk analysis in general (Antonioni et al., 2007; Krausmann et al., 2011a):

1. natural-hazard severity parameters [e.g., peak ground acceleration (PGA), which is the most commonly used proxy for describing an earthquake's severity and damage potential, peak ground velocity, peak ground displacement, etc.];
2. target equipment (usually the highest priority would be given to the most dangerous equipment categories both in terms of type and quantity of hazardous substances processed or stored and the equipment's operating conditions);
3. damage states (the earthquake severity needs to be related to the damage intensity; this can be achieved by lessons-learned type studies of past accidents or using numerical modeling);
4. equipment damage models (probit models or fragility curves which relate the damage intensity to the associated probability);
5. consequence-analysis models [these models estimate the consequences of a LOC, e.g., in terms of substance concentrations (toxic release), heat radiation (fires), or overpressure (explosions)];
6. likelihood estimates (frequencies, probabilities, or qualitative likelihood estimates for all possible event combinations);
7. information on risk receptors (e.g., population distribution around the hazardous installations).

Empirical equipment damage models were developed or are under development, partly based on the analysis of past accident data (Salzano et al., 2003; Campedel et al., 2008; Antonioni et al., 2009). The lack of detailed equipment damage models is the main limitation of the methodology and a significant source of uncertainty. Further work in this direction is therefore required. This will also decrease data and model uncertainties inherent in the analysis.

Some natural hazards, for example, a strong earthquake or flood causing a Natech accident are likely to simultaneously down on- and off-site lifelines and utilities required for accident prevention and mitigation. The loss of utilities can either trigger the Natech accident in the first place, or hamper emergency-response actions to mitigate its consequences. Therefore, a worst-case risk-analysis approach seems warranted in which the failure of internal and external safety and mitigation measures is assumed in the QRA scenario-building process.

7.3.2 Hazard Identification and Consequence Analysis

The general framework for industrial risk assessment outlined in Section 7.2 is equally valid for Natech risk assessment. However, the conventional QRA procedure needs to be extended to accommodate the characteristics of Natech accident scenarios. Hence, specific damage models to assess the severity and probability of equipment damage due to a natural event, and a procedure to account for the possibility of simultaneous hazardous-materials releases from more than a single process or storage unit are required (Krausmann et al., 2011a).

The flowchart in Fig. 7.3, which was adapted from Antonioni et al. (2007), shows the principal steps in the quantitative analysis of seismic Natech risk. This

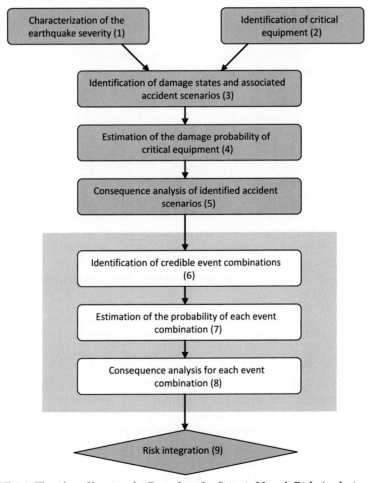

FIGURE 7.3 Flowchart Showing the Procedure for Seismic Natech Risk Analysis

Adapted from Antonioni et al. (2007).

methodology is also applicable to the analysis of Natech risk due to any other type of natural-event trigger. The starting point in the analysis of seismic Natech risk is the characterization of the natural hazard in terms of severity and frequency at the location of the hazardous installation (Step 1) and the identification of the target equipment (Step 2). Information on the PGA is easily available from measurements or historical records, and PGA is therefore practical for characterizing the severity of the ground motion (Campedel et al., 2008). Probabilistic seismic hazard analysis (PSHA) can be used to estimate the occurrence probability of an earthquake with a given intensity at a specific location. For the analysis of flood Natech risk, information from flood hazard maps prepared by government authorities, or reports and anecdotal information on past inundations could be used to characterize the flood severity in terms of water height, speed, and duration (Krausmann, 2016).

The natural-hazard scenarios on which the risk analysis is to be based are often determined by the regulator which may require protection of a hazardous installation against specific natural-hazard levels (e.g., a 100-year flood or an earthquake with 475 years return period). The operator can decide to voluntarily ramp up safety levels above minimum requirements, and operators that have already been affected, for example, by floods, recommend using worst-case natural-hazard scenarios in the site's external hazard identification.

The target equipment that poses the greatest safety risk can be identified based on evidence of its involvement in past Natech accidents, determined from chemical-accident database analyses (Sengul, 2005; Krausmann et al., 2011b; Santella et al., 2011) and the criticality in case of LOC. The latter is defined by the equipment's operating conditions in terms of temperature and pressure and the amount and type of substance it contains. These factors, together with the extent of damage suffered by a piece of equipment due to earthquake loading or impact by another type of natural hazard, also influence the accident scenarios which can develop following a LOC event.

Once the target equipment has been identified, the possible damage caused by a natural hazard has to be associated with it. Detailed earthquake damage models are available for certain types of equipment, for example, storage tanks, from structural-engineering applications. Use of these models is, however, not practicable for assessing the risk to a chemical complex with its many units and structures. Following the approach used in HAZUS (FEMA, 2003), the damage to a piece of equipment can be approximated by defining a limited number of discrete damage states DS (Step 3). These damage states are then used to calculate the expected severity of LOC. Depending on the application, a different number of damage states can be applied. Antonioni et al. (2007) use two damage states, where DS1 indicates limited structural damage with a partial loss of vessel inventory or entire loss in a time interval higher than 10 min, and DS2 indicates extended structural damage with complete loss of inventory in less than 10 min. On the other hand, Salzano et al. (2003) define three damage states with DS1 and DS2 denoting minor and relevant structural damage, respectively, and DS3 corresponding to total structural collapse. In order to describe the severity of release they then associate three so-called risk states with these damage states (minor and relevant release, loss of complete inventory). Other damage and risk states can be defined.

The accident scenario(s) following LOC depend on the substance hazard and the extent of release, which is defined by the magnitude of the LOC (small or big hole) and the storage or operating conditions. For instance, even if the damage severity is the same, releases from pressurized equipment could have very different consequences than discharges from vessels kept at atmospheric pressure. The quantity and conditions of the released substance can be determined using source-term models. Once the LOC severity is known, event trees can be applied to identify the final outcomes of the scenarios of a specific damage mode (e.g., pool fire, jet fire, toxic dispersion, vapor cloud explosion, etc.). Specific event trees, derived from the analysis of past Natech accidents, should be used to consider the dynamics of Natech events (Cozzani et al., 2010; Renni et al., 2010). Although a realistic Natech risk analysis would also require an assessment of the natural-hazard impact on existing protection measures, Antonioni et al. (2007) recommend disregarding prevention and mitigation measures as they would likely be damaged by the earthquake and therefore not be available.

The probability of a target equipment to suffer a specific extent of damage as function of the earthquake severity (e.g., PGA) can be estimated from fragility curves or using so-called probit functions (Finney, 1971; Vílchez et al., 2001), with both approaches being equivalent (Step 4). Although some fragility information for typical chemical storage and processing equipment is available (e.g., Kiremidjian et al., 1985; Seligson et al., 1996; Salzano et al., 2003; Fabbrocino et al., 2005; Di Carluccio et al., 2006), probit functions are more widely used in industrial quantitative risk analysis. The probit variable Pr_{DS} is a function of the PGA and can be easily converted into the equipment damage probability P_{DS} for a specific damage state DS (Finney, 1971):

$$Pr_{DS} = k1 + k2\ln(PGA)$$

The frequency of the accident scenario associated with specific equipment damage is then calculated by multiplying the frequency of an earthquake of a given PGA value with P_{DS}. Campedel et al. (2008) provide a list of the probit coefficients $k1$ and $k2$ for different types of equipment based on empirical data. The probit coefficients also depend on the existence of implemented safety measures (e.g., anchoring of equipment) and the filling level which determines the criticality of liquid-sloshing effects due to the earthquake forces (Ibrahim, 2005). The lack of detailed equipment damage models linking damage states to probabilities is at present one of the main limitations of the proposed analysis methodology.

The consequences of the final outcomes of an accident scenario can be assessed using conventional consequence models (Step 5). They include not only dispersion, fire, and explosion models, but also effects models including probit functions that allow the calculation of the impact of overpressure, heat radiation, and toxic effects on the population (e.g., van den Bosch and Weterings, 1997; CCPS, 2000). A brief synopsis of consequence-analysis models is provided in Section 7.3.2.1.

For some Natech accidents there is a high likelihood that several chemical process or storage units are damaged simultaneously by the natural hazard and hence more than one accident scenario will occur. This is taken into account in Steps 6–8 where the event combinations are identified, and their frequencies and consequences evaluated. The overall consequences of the combined scenarios can then be derived by summing up the results, for example, human health impact for each single accident scenario. The calculation of the overall frequency of the identified combinations of scenarios is presented in detail in Antonioni et al. (2007). In the last step (9) the risk from the identified accident scenarios is estimated and visualized as individual risk or F–N curves.

7.3.2.1 Consequence-Analysis Models

Although it is beyond the scope of this book to provide a detailed discussion of consequence-analysis models, a brief introduction to the subject is beneficial for the overall understanding of the risk-analysis process. Christou (1998b) provides a generic but concise overview of the most common phenomena and the associated models used in the analysis. van den Bosch and Weterings (1997) give a more detailed description of available models and the conditions under which they should be used.

The analysis of the consequences of a loss-of-containment event crucially depends on the *source term*, that is, the quantity and the conditions of the hazardous-materials release. Discharge models are readily available and provide the amount and rate of substance outflow, the duration of the discharge, and the phase of the release (liquid, gas, or two-phase).

If the substance is discharged into the atmosphere, *dispersion models* are applied to model the behavior of the formed substance cloud. Depending on the properties of the released gas (lighter or heavier than air), specific models are used. For atmospheric dispersion modeling, meteorological conditions at the time of substance release (wind speed and direction, temperature, humidity, pressure, and air stability class) and the site topography are relevant. Dispersion models provide information on the substance concentration as a function of location and time.

When flammable substances are released there is a risk of ignition. Depending on the process or storage conditions, *pool fires* (substances at atmospheric pressure) or *jet fires* (pressurized substances) can result. The associated fire models calculate the thermal flux (heat radiation) emanating from the fires. If a fire ignites close to pressurized equipment, the heating of the substance and the subsequent increase in pressure inside the unit can cause a mechanical explosion, destroying the unit. This *boiling liquid expanding vapor explosion* (BLEVE) releases the full substance content of the affected unit instantaneously. Of importance for analyzing the consequences of a BLEVE are the formation of a fireball (thermal radiation) and the generation of projectiles. An *unconfined vapor cloud explosion* (UVCE) is usually the product of delayed ignition of a discharged flammable substance. If a flammable vapor cloud finds no immediate ignition source upon release, the cloud is dispersed with the wind and can ignite at a later stage. Of concern with respect to UVCEs is the formation of a rapidly moving pressure wave and the associated effects of overpressure on risk receptors.

In industrial risk assessment, vulnerability models aim to estimate the response of risk receptors (e.g., in terms of health effects) to toxic concentration, heat radiation, and overpressure from shock waves. Due to the different response behavior of individuals in particular to toxic exposure, vulnerability assessment is affected by significant uncertainty. Toxic-effects models are based on laboratory experiments which are then extrapolated to humans (definition of threshold concentration or dose for death and injury). For the impact of heat from fires, radiation intensities and the associated expected effects on the population and structures are defined [e.g., 37.5 kW/m^2 is sufficient to cause damage to process equipment (Christou, 1998b)]. For explosions, shock wave overpressure models are used. Probit functions also lend themselves to the assessment of all aforementioned effects.

7.3.2.2 Cascading Effects

Industrial installations are often grouped into large integrated clusters due to economic reasons, environmental factors, legal requirements, or social issues. Every installation in such an industrial park represents a danger to the other facilities in the neighborhood depending on the amount and type of hazardous substances processed, handled, or stored, and the associated process or storage conditions. If an accident occurs in one of these installations there is therefore the risk of a domino or cascading effect, that is, a propagation of the accident within the same plant or to one or more neighboring installations due to knock-on effects. This could lead to an escalation of the primary accident with consequences on potentially extended areas of the industrial park (Reniers and Cozzani, 2013). For Natech accidents, the risk of a cascading effect is particularly high due to the potentially multiple and simultaneous hazardous-materials releases over extended areas, and the loss of lifelines needed for accident prevention and mitigation (Chapter 3).

The direct causes of cascading or domino effects are blast waves, fires, or the projection of fragments caused by a primary accident with equipment damage and the subsequent LOC of hazardous substances. Cozzani et al. (2013) argue that also indirect effects can lead to a domino accident. The release of hazardous materials would in this case not be triggered by direct equipment damage but rather via, for example, loss of process control due to damage to the control room, or structural damage to and collapse of buildings housing hazardous equipment. Loss of cooling or power fluctuations can also lead to process upsets and LOC (Section 2.6).

A significant number of domino accidents occurred in the past and with their consequences being severe, European legislation on the control of major accident hazards involving dangerous substances explicitly addresses this type of risk. The Seveso Directive requires the identification, assessment, and control of domino risks for all establishments covered by the Directive (Section 4.1.1). This also includes the requirement that concerned operators exchange suitable information that allows them to take appropriate measures to control the domino risk. Appropriate safety distances between hazardous equipment, thermal insulation of units to protect them against heat impingement from adjacent fires, or emergency water deluges are some preventive measures that can be taken to address domino risks (Mecklenburgh, 1985). Consolidated methodologies and tools for the identification of domino risks and for GIS-supported quantitative domino risk assessment are available (Khan and Abbasi, 1998, 2001; Cozzani et al., 2005, 2006).

For an in-depth treatment of the identification and assessment of domino scenarios in the process industries, both from a safety and security context, the reader is referred to Reniers and Cozzani (2013).

7.3.3 Risk Integration and Evaluation

For risk integration the most commonly used indicators are individual risk and societal risk. The numerical outcome of a quantitative risk analysis is not an accurate number but invariably contains uncertainties introduced during the various steps of the analysis process which can be up to one order of magnitude. The sources of these uncertainties can be grouped into uncertainties in models, input data and in general analysis quality, and range from outdated failure rates to errors in consequence modeling and omissions in the identification and characterization of all relevant hazards (Cox, 1998; CCPS, 2000). They do not render the analysis and its results invalid. However, end users have to be aware of the existence of uncertainties when making decisions based on risk estimates. CCPS (2000) provides an exhaustive list of sources of uncertainty in the risk-analysis process.

Risk analysis can be used in several ways to improve the safety of a hazardous installation. It helps identify system weaknesses, provides input for the setting of risk-reduction priorities, or quantifies the improvement achieved by the implementation of targeted risk-reduction measures for optimizing expenditures. Alternatively, the outcome of risk analysis feeds into the decision-making process by providing risk estimates that can be used for comparison with risk criteria that need to be complied with to fulfill regulatory requirements. This step in the risk-assessment process is also referred to as risk evaluation.

Although the analysis of risk is a rather scientific task, the management of the resulting risk involves a great deal of consultation and negotiation between all stakeholders. This includes decisions on which types of risk and associated risk levels are acceptable or tolerable and which are not. In this context, acceptable and tolerable are not identical. Instead, "tolerable" refers to the willingness of society to tolerate certain risks as long as the benefit outweighs the risks (HSE, 2001). Moreover, in most legal frameworks for chemical-accident prevention proof needs to be given that the risk associated with a hazardous activity is controlled. This includes the identification and implementation of adequate risk-reduction measures.

There are several methods to judge whether a specific risk level is sufficiently low. A cost-benefit analysis expresses the residual risk in terms of monetary value. The term "as low as reasonably practicable (ALARP)" defined in the United Kingdom indicates an approach that avoids risk reduction that would incur grossly disproportionate costs (Cox, 1998). Other possibilities to judge whether a risk is acceptable are expert judgment or prescriptive numerical risk targets or criteria. In addition to

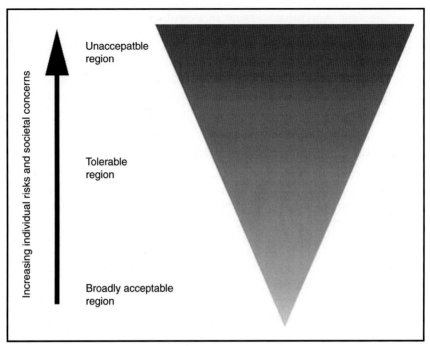

FIGURE 7.4 Carrot Diagram Representing the Framework for Risk Acceptance (HSE, 2001)

the regulatory regime and the political context, the definition of these absolute risk criteria strongly depends on society's attitude toward hazards.

Practically, there is a general framework that defines three levels of risk as illustrated in Fig. 7.4 (HSE, 2001). In the top zone risk is unacceptable and must be reduced at any cost. Risks falling into the zone at the bottom are considered negligible and usually no further risk-reduction actions are required. The tolerable risk zone would then lie in between these boundary levels. This area is subject to proper risk assessment and the implementation of risk-control measures to achieve risk levels that are as low as practicable.

In the European Union (EU) there are no uniform risk criteria as safety approaches (probabilistic, deterministic) vary across its Member States. Some countries have developed risk-based criteria, defining acceptable levels of risk, while others specify consequence-based criteria which establish maximum permissible levels of overpressure, heat radiation, or toxic concentration. Based on the risk criteria used in selected EU Member States, Trbojevic (2005) suggests a set of common criteria for individual risk. In his proposal the boundary for unacceptable individual risk is set to 10^{-5} events per year while negligible risk is 10^{-8} per year. The target level for individual risk within the tolerable risk zone should be 10^{-6} per year.

The decision about the acceptability of risk must take both individual risk and societal concerns into account. Developing criteria for the risk of multiple fatalities in one accident is, however, difficult and only few countries have defined numerical values which are either prescriptive or used as nonlegal orientation norms. As an example, in the Netherlands the expected frequency of accidents involving more than 10 fatalities must not exceed 10^{-5} per year, while for accidents with more than 100 deaths the frequency must not be greater than 10^{-7} per year (Duijm, 2009).

References

Antonioni, G., Bonvicini, S., Spadoni, G., Cozzani, V., 2009. Development of a framework for the risk assessment of Natech accidental events. Reliab. Eng. Syst. Saf. 94, 1442–1450.

Antonioni, G., Spadoni, G., Cozzani, V., 2007. A methodology for the quantitative risk assessment of major accidents triggered by seismic events. J. Hazard. Mater. 147, 48–50.

Campedel, M., Cozzani, V., Garcia-Agreda, A., Salzano, E., 2008. Extending the quantitative assessment of industrial risks to earthquake effects. Risk Anal. 28 (5), 1231–1246.

CCPS, 2000. Guidelines for Chemical Process Quantitative Risk Analysis, second ed. Center for Chemical Process Safety/American Institute of Chemical Engineers, New York.

Christou, M.D., 1998a. Introduction to risk concepts. Kirchsteiger, C., Christou, M.D., Papadakis, G.A. (Eds.), Risk Assessment and Management in the Context of the Seveso II Directive, Industrial Safety Series, vol. 6, Elsevier, Amsterdam.

Christou, M.D., 1998b. Consequence analysis and modelling. Kirchsteiger, C., Christou, M.D., Papadakis, G.A. (Eds.), Risk Assessment and Management in the Context of the Seveso II Directive, Industrial Safety Series, vol. 6, Elsevier, Amsterdam.

Cox, T., 1998. Risk integration and decision making. Kirchsteiger, D., Christou, M.D., Papadakis, G.A. (Eds.), Risk Assessment and Management in the Context of the Seveso II Directive, Industrial Safety Series, vol. 6, Elsevier, Amsterdam.

Cozzani, V., Campedel, M., Renni, E., Krausmann, E., 2010. Industrial accidents triggered by flood events: analysis of past accidents. J. Hazard. Mater. 175, 501.

Cozzani, V., Krausmann, E., Reniers, G., 2013. Other causes of escalation. In: Reniers, G., Cozzani, V. (Eds.), Domino Effects in the Process Industries. Elsevier, Waltham, MA.

Cozzani, V., Gubinelli, G., Antonioni, G., Spadoni, G., Zanelli, S., 2005. The assessment of risk caused by domino effect in quantitative area risk analysis. J. Hazard. Mater. 127, 14.

Cozzani, V., Gubinelli, G., Salzano, E., 2006. Escalation thresholds in the assessment of domino accidental events. J. Hazard. Mater. 129 (1–3), 1–21.

Cruz, A.M., Okada, N., 2008. Consideration of natural hazards in the design and risk management of chemical industrial facilities. Nat. Hazard. 44 (2), 213–227.

Delvosalle, C., Fievez, C., Pipart, A., 2004. ARAMIS: accidental risk assessment methodology for industries in the context of the Seveso II Directive. ARAMIS project deliverable D.1.C., July 2004.

Di Carluccio, A., Iervolino, I., Manfredi, G., Fabbrocino, G., Salzano, E., 2006. Quantitative probabilistic seismic risk analysis of storage facilities. In: Proc. CISAP-2, Chem. Eng. Trans. 9.

Duijm, N.J., 2009. Acceptance criteria in Denmark and the EU. Danish Ministry of the Environment, Environmental Protection Agency, Environmental Project No. 1269.

Fabbrocino, G., Iervolino, I., Orlando, F., Salzano, E., 2005. Quantitative risk analysis of oil storage facilities in seismic areas. J. Hazard. Mater. A123, 61–69.

FEMA, 2003. Multi-hazard Loss Estimation Methodology, Earthquake Model, HAZUS®MH MR4, Technical Manual, US Federal Emergency Management Agency, Washington D.C.

Finney, D.J., 1971. Probit Analysis. Cambridge University Press, UK.

HSE, 2001. Reducing risks, protecting people—HSE's decision-making process. Health and Safety Executive, United Kingdom.

HSE, 2009. Societal risk: initial briefing to Societal Risk Technical Advisory Group. Research Report RR703, Health and Safety Executive, United Kingdom.

Ibrahim, R.A., 2005. Liquid Sloshing Dynamics: Theory and Applications. Cambridge University Press, Cambridge.

Khan, F.I., Abbasi, S.A., 1998. Models for domino effect analysis in chemical process industries. Process Saf. Prog. 17, 107.

Khan, F.I., Abbasi, S.A., 2001. An assessment of the likelihood of occurrence, and the damage potential of domino effect (chain of accidents) in a typical cluster of industries. J. Loss Prevent. Proc. Indus. 14 (4), 283.

Kiremidjian, A., Ortiz, K., Nielsen, R., Safavi, B., 1985. Seismic risk to industrial facilities. Report No. 72, The John A. Blume Earthquake Engineering Center, Stanford University.

Krausmann, E., 2016. Natech accidents—an overlooked type of risk? Loss Prev. Bull. 251, October 2016, Institution of Chemical Engineers, United Kingdom.

Krausmann, E., Cozzani, V., Salzano, E., Renni, E., 2011a. Industrial accidents triggered by natural hazards: an emerging risk issue. Nat. Hazard. Earth Syst. Sci. 11, 921–929.

Krausmann, E., Renni, E., Campedel, M., Cozzani, V., 2011b. Industrial accidents triggered by earthquakes, floods and lightning: lessons learned from a database analysis. Nat. Hazard. 59/1, 285–300.

Mecklenburgh, J.C. (Ed.), 1985. Process plant layout, Goodwin in association with the Institution of Chemical Engineers.

Reniers, G., Cozzani, V. (Eds.), 2013. Domino Effects in the Process Industries. Elsevier, Waltham, MA.

Renni, E., Krausmann, E., Cozzani, V., 2010. Industrial accidents triggered by lightning. J. Hazard. Mater. 184, 42.

Salzano, E., Iervolino, I., Fabbrocino, G., 2003. Seismic risk of atmospheric storage tanks in the framework of quantitative risk analysis. J. Loss Prevent. Proc. Indus. 16, 403–409.

Santella, N., Steinberg, L.J., Aguirre, G.A., 2011. Empirical estimation of the conditional probability of Natech events within the United States. Risk Anal. 31 (6), 951–968.

Seligson, H.A., Eguchi, R.T., Tierney, K.J., Richmond, K., 1996. Chemical hazards, mitigation and preparedness in areas of high seismic risk: a methodology for estimating the risk of post-earthquake hazardous materials releases. Technical Report NCEER-96-0013, National Center for Earthquake Engineering Research, State University of New York at Buffalo.

Sengul, H., 2005 Hazard characteriztion of joint natural and technological disasters (Natechs) in the United States using federal databases. Master Thesis, Dept. of Civil and Environmental Engineering, Tulane University, New Orleans, USA.

Trbojevic, V.M., 2005. Risk criteria in EU. In: Proc. ESREL 2005, Tri-City, Poland, 27–30 June.

Uijt de Haag, P.A.M., Ale, B.J.M., 1999. Guidelines for quantitative risk assessment (Purple Book), CPR 18E, Committee for the Prevention of Disasters, The Hague, The Netherlands.

van den Bosch, C.J.H., Weterings, R.A.P.M., 1997. Methods for the calculation of physical effects (Yellow Book), Committee for the Prevention of Disasters, The Hague, The Netherlands.

Vílchez, J.A., Montiel, H., Casal, J., Arnaldos, J., 2001. Analytical expressions for the calculation of damage percentage using the probit methodology. J. Loss Prevent. Process Indus. 14, 193–197.

Qualitative and Semiquantitative Methods for Natech Risk Assessment

E. Krausmann*, K.-E. Köppke**, R. Fendler†, A.M. Cruz‡, S. Girgin*
*European Commission, Joint Research Centre, Ispra, Italy; **Consulting Engineers Prof. Dr. Köppke, Bad Oeynhausen, Germany; †German Federal Environment Agency, Dessau-Roßlau, Germany; ‡Disaster Prevention Research Institute, Kyoto University, Kyoto, Japan

Risk assessment is a prerequisite for understanding the Natech risk and for determining if and which prevention and preparedness measures should be implemented to reduce the risk. The analysis of multihazard risks is a highly complex task and there is no consolidated methodology for assessing the Natech risk. This chapter introduces selected qualitative and semiquantitative Natech risk-analysis methodologies, approaches, and tools of varying levels of resolution. The outcome of these methodologies can be used for evaluating the risk in accordance with the risk-acceptability criteria in place.

8.1 RAPID-N

The identification of potentially Natech-prone areas and the determination of the associated risk level are the first steps toward managing Natech risks. As Krausmann and Baranzini (2012) note, hardly any Natech risk maps exist in EU Member States and OECD Member Countries. In few countries, maps with overlays of natural and technological hazards were created to provide an indication to government authorities of where multihazard risks could exist. These hazard maps do, however, not consider site-specific features or the interaction between natural and technological hazards. The development of a Natech risk assessment and mapping capability is considered a high-priority need by authorities for effectively reducing Natech risks.

Following calls by government, the European Commission's Joint Research Centre has developed a semiquantitative methodology for Natech risk analysis and mapping which has been implemented as a web-based software framework called RAPID-N. It is freely available via prior user registration and authorization at http://rapidn.jrc.ec.europa.eu. RAPID-N allows the quick analysis of Natech risks at local (single installation) or regional (multiple assets) level with a minimum of data. It features a user-friendly interface with advanced data entry, visualization, and analysis tools. In order to preserve confidentiality, it supports data protection and access restrictions for critical information, such as industrial plant data and the associated risk assessment.

RAPID-N was designed to support different natural hazards and industrial equipment types. Estimating the Natech risk and mapping it in a web-based environment, it can not only support land-use- and emergency planning, but also Natech damage and consequence analysis immediately after a natural event. The latter is fundamental for first responders who require an assessment of the dangers of secondary hazards from industrial plants following a natural disaster before dispatching rescue teams. It could also provide a means for authorities to warn the population in the vicinity of the installation in a timely manner.

The structure of RAPID-N is based on four self-contained but interconnected modules, each of which carries out specific tasks in support of the Natech risk-analysis process. These modules are briefly described in the following sections. For a detailed description of the framework and guidance on its use, the reader is referred to Girgin and Krausmann (2013) and Girgin (2012). In Chapter 10 a full case-study application of RAPID-N is presented for earthquake impact on a chemical installation containing flammable and toxic substances.

8.1.1 Scientific Module

The scientific module provides all underlying computational support for the RAPID-N simulations (data handling, statistics, mapping, etc.). The module also includes the so-called *property-definition and -estimation framework*, which forms the very basis of RAPID-N's Natech risk-analysis functionality. Natural hazards, industrial plants and their units, and hazardous substances are all described via their properties, for example, hazard severity, site characteristics, boiling point, etc. These characteristics are not always known in sufficient detail for RAPID-N to run a simulation. Hence, the tool uses property estimators to fill existing data gaps based on available scientific estimation methods and equations. For example, RAPID-N can estimate the peak ground acceleration at the hazardous installation based on the epicentral severity and the use of automatically selected attenuation equations valid for the given geographic region. An example of a very simple property estimator would be the automatic calculation of the tank diameter from its volume. Instead of analytical equations or complex mathematical functions, default values of the properties can also be specified (e.g., the default ambient temperature). Rather than using hard-coded functions, RAPID-N also performs the damage and risk analysis with property estimators. The advantage of this implementation is twofold: (1) it reduces the amount of data to be provided by the user, and (2) the Natech risk analysis is rendered extremely flexible as alternative assessment methods with varying complexity can be implemented easily by the user. The estimation framework also allows the definition of custom properties to support additional analysis needs. Uncertainty in numerical data can be specified using fuzzy numbers, which are automatically taken into consideration by RAPID-N while performing numerical calculations. Wherever possible, the calculation results are also reported as fuzzy numbers to provide an indication of the uncertainty.

8.1.2 Industrial Plants and Units Module

For assessing the Natech risk, RAPID-N estimates the damage severity and probability for industrial-plant equipment under natural-hazard loading. For this purpose, the tool needs information on the geographic location of the plant and the characteristics of the plant units, such as type of equipment, dimensions, structural properties, operating and storage conditions, and hazardous substances contained in the unit. Existing safety measures, for example, dikes around storage tanks, can also be specified. RAPID-N includes a special mapping tool that helps the user to quickly locate and delineate the boundaries of industrial installations and identify plant units using publicly available satellite imagery. Once the type and total quantity of hazardous substance has been defined, RAPID-N calculates the amount of substance involved in the accident and used for the consequence analysis based on operating/storage conditions. Fig. 8.1 gives an example of a RAPID-N interface showing both user-defined and estimated plant unit information. Process units located at fixed chemical installations and onshore pipelines are currently supported.

In-depth information on plant and equipment characteristics is usually difficult to obtain as it is often considered proprietary and is therefore closely held by industry. If no equipment information is available at all, user-defined *typical plant units* can be assigned to a hazardous plant to allow a rough estimate of the Natech risk. These typical plant units can be defined for specific industrial activities, for example, typical storage tanks for refineries or petrochemical facilities.

8.1.3 Natural-Hazards Module

Source and onsite data on the natural-hazard accident initiator are provided by RAPID-N's natural-hazards module. The source data include information on the characteristics of the natural hazard, for example, earthquake coordinates, focal depth, and magnitude. However, for the Natech risk analysis the source characteristics are secondary. Rather, the natural-hazard severity data at the location of the hazardous installation are required. RAPID-N can estimate the onsite natural-hazard severity by using estimation equations that are selected automatically by the data-estimation framework based on source parameters, location, and geographic region. For example, RAPID-N uses attenuation equations for estimating the seismic forces at a specific distance from the epicenter. RAPID-N also supports the direct inputting of the onsite natural-hazard severity, if known to the user, or the use of earthquake hazard maps (e.g., USGS ShakeMaps) for which local hazard parameters are calculated by interpolation of available map data.

RAPID-N currently contains data and hazard-intensity estimation methods that focus on earthquakes as the proof-of-concept hazard. Its natural-hazard

Process Unit Information

Facility:	Turkish Petroleum Refineries Corp. (TUPRAS) Izmit Refinery, Turkey
Type:	Storage Tank
Coordinate:	44' 51.515", 46' 15.131"
Substance:	Naphtha (8030-30-6)

Properties

Shape:	Cylindrical Vertical
Volume:	31979 m³*
Height:	13.46 m*
Diameter:	55 m
H/D Ratio:	0.2447 m/m*
Roof Type:	Floating Roof
Base Type:	On-ground*
Base Support Type:	Unanchored*
Construction Material:	Steel
Fill Level:	85 %v*
Fill Height:	11.441 m*
Stored Volume:	27182 m³*
Storage Volume:	31979 m³*
Storage Condition:	Atmospheric
Storage Pressure:	1 atm*
Storage Temperature:	25°C*
Passive Release Mitigation Factor:	1*

FIGURE 8.1 Example of Plant Unit Information

The asterisk designates data estimated or assigned as default values by RAPID-N in the absence of user-defined information.

database includes source data of worldwide earthquakes with a magnitude >5.5 since the early 1970s for use in historical scenario assessment. RAPID-N also monitors the earthquake catalogs of the United States Geological Survey (USGS) and the European-Mediterranean Seismological Centre (EMSC), and once new earthquake data become available they are automatically included in RAPID-N's earthquake database. At the time of writing this book, work has been launched to provide data and methods for extending the tool to flood Natech risk analysis and mapping.

8.1.4 Natech Risk-Analysis Module

This module constitutes the very core of RAPID-N. It estimates equipment damage caused by a natural hazard, analyzes the most likely consequences, and visualizes the outcome of the analyses on a map showing possible impact zones.

The analysis process starts with the determination of the damage probability and severity of equipment located at industrial plants impacted by a natural hazard. For this purpose, RAPID-N uses fragility curves which relate onsite hazard-severity parameters (e.g., peak ground acceleration) to the probability of damage for a given damage severity (Fig. 8.2). In order to be able to define damage-state-specific Natech scenarios, RAPID-N uses discrete damage states, examples of which are given in Section 7.3 and in FEMA (2003). Since criteria for damage classification also depend on the target application area (e.g., analysis of hazardous-materials releases vs. economic cost of damage) RAPID-N supports multiple damage classifications to allow the user to customize the risk analysis. If the user prefers, RAPID-N can automatically select for each plant unit an appropriate fragility curve.

Once the level of damage is known, it needs to be related to type and severity of loss of containment to identify the consequences in terms of toxic dispersion, fire, or explosion. RAPID-N associates risk states to the predicted damage severities which in practice are simplified consequence scenarios. These risk states define the magnitude of loss of containment, depending on the substance hazard and the operating or storage conditions, and the release and ignition probabilities where applicable. Multiple risk states with different scenario parameters and validity conditions can be defined for a specific damage state. The most appropriate risk state for further analysis is automatically determined by RAPID-N considering plant unit characteristics, damage state, and validity conditions of the risk states.

Based on the selected risk state, RAPID-N calculates the amount of substance involved in the accident, which is used for the consequence analysis depending on operating/storage conditions. The consequences are estimated by means of a scenario-specific, dynamically-generated consequence model that is formed by RAPID-N using the consequence-analysis methods and equations available in the database of the property-estimation framework. As proof of concept, the framework includes the US EPA RMP Guidance for Offsite Consequence Analysis methodology (US EPA, 1999) for analyzing the impacts of the probable Natech accident scenarios. While it is not a full-fledged quantitative analysis methodology, it is an example of a simple but functional approach to consequence analysis that allows the estimation of the distances to severity endpoints. For toxic releases, the endpoints are either Emergency Response Planning Guideline 2 (ERPG-2) or immediately dangerous to life and health (IDLH) toxic concentrations. For flammable substances, the endpoints to heat radiation levels of 5 kW/m^2 for 40 s (corresponding to second degree burns) and overpressures of 7 kPa (1 psi) for vapor cloud explosions are calculated. The user can easily customize the endpoint criteria or introduce different consequence-analysis models, if so preferred. User-defined modifications to the RAPID-N framework do not affect other users, leaving room for experimenting with novel Natech risk-analysis approaches.

Fragility Curve Information

Name:	HAZUS, On-ground anchored steel tank
Abbreviation:	HAZUS
Process Unit Type:	Storage Tank
Damage Classification:	HAZUS (2010)
Hazard Parameter:	Peak Ground Acceleration (PGA)
Unit:	g
Functional Form:	Log-normal (median)
Precedence:	Auto

Validity Conditions

Base Type:	On-ground
Base Support Type:	Anchored
Construction Material:	Steel

Data

No	Damage State	Median	Standard Deviation
1.	≥ DS2	0.3	0.6
2.	≥ DS3	0.7	0.6
3.	≥ DS4	1.25	0.65
4.	= DS5	1.6	0.6

References

No	Reference
1.	U.S. EPA, "HAZUS-MH MR5 Technical Manual", 2010

Fragility Curve

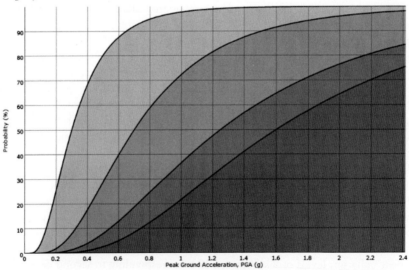

FIGURE 8.2 Fragility Curve for an Anchored Steel Storage Tank Subjected to Earthquake Shaking

The RAPID-N output is presented as a summary report including all parameters used in the risk analysis and detailed results, as well as an interactive risk map that displays the severity, probabilities, and impact zones of the simulated Natech events. Fig. 8.3 gives an example of a case-study output. If the risk assessment involves multiple plant units, areas which might be affected by releases from several units can

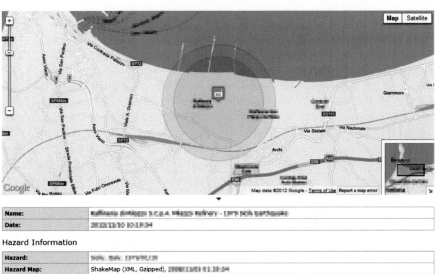

Name:	████████ ████████ S.C.p.A. ████████ ████████ - ████ ████ ████████
Date:	████████ ██████

Hazard Information

Hazard:	████, ████, ████████
Hazard Map:	ShakeMap (XML, Gzipped), ████████ ██████

Industrial Plant Information

Industrial Plant:	████████ ████████ S.C.p.A. ████████ ████████, ████
Plant Unit:	Storage Tank, 96, Propane (74-98-6)

Damage Estimation

Damage Classification:	Auto
Flexible fragility curve selection:	No

Industrial Plants

1. ████████ ████████ S.C.p.A. ████████ ████████, Italy

No	Plant Unit	Hazard Parameters	Fragility Curve	Damage Estimate	Damage Parameters	End-point Distance
1.	Storage Tank (96) Propane, Q_{stored}: 878456 kg; V: 1767.1 m³; Shape: Spherical; d: 15 m; r: 7.5 m; φ_{hd}: 1 m/m; d_b: 7.5 m; Base Type: Above Ground; $V_{storage}$: 1767.1 m³; $Q_{storage}$: 1033478 kg; Storage Condition: Pressure; $T_{storage}$: 25°C; Storage State: Liquid; Fill Percent: 85 %v; h_{fill}: 11.334 m; V_{stored}: 1502.1 m³; $f_{m, passive}$: 1; $f_{m, active}$: 1 ≪	PGA: 9.9962 %g; d_e: 23.384 km; MM: Rather Strong; MSK: Fairly strong; EMS: Strong; MMI: 5.4687; PGA_h: 41.418 cm/s²; PGV: 6.3253 cm/s ≪	OS00-F50-G	≥ DS2: 0.1896% ≥ DS3: 1.6883·10⁻⁶%	Fire/Explosion Event: Vapor Cloud Explosion; $Q_{released}$: 87846 kg ≫ Fire/Explosion Event: Vapor Cloud Explosion; $Q_{released}$: 175691 kg; $f_{V, involved}$: 20 %v; $f_{Q, involved}$: 20%; $V_{involved}$: 300.41 m³; $Q_{involved}$: 175691 kg; P_{damage}: 1.6883·10⁻⁸; $P_{c, release}$: 50%; Release State: Gas; $q_{release}$: 17569 kg/min; $t_{release}$: 10 min; $T_{release}$: 25°C; $V_{released}$: 300.41 m³; $h_{release}$: 0 m; q_{gas}: 17569 kg/min; $q_{gas, reduced}$: 17569 kg/min; t_{gas}: 10 min; PT: Dense; P_{natech}: 1.6883·10⁻⁶%; $P_{c, fire}$: 100%; RT_{RMP}: Table 7; Q_{fuel}: 175691 kg; Δp: 1 psi; f_{yield}: 0.1; d_e: 948.96 m ≪	753.2 m: 0.1896% 949 m: 1.6883·10⁻⁶%

FIGURE 8.3 Example Output of a RAPID-N Earthquake Natech Case Study

be easily identified in the output map. Moreover, as the risk of cascading effects during Natech events is high, RAPID-N can also be used as a screening tool for identifying potential problem areas due to domino effects. For example, if released flammable substances ignite, RAPID-N shows if other infrastructures fall within the fire's impact zone. This gives an indication of where attention should be paid and where further in-depth analysis might be warranted.

8.1.5 Outlook

The current version of RAPID-N supports earthquake Natech risk analysis and mapping for fixed chemical installations and onshore pipeline systems. The next release of the tool will incorporate floods as additional Natech accident trigger. RAPID-N will then also be able to calculate individual and societal risk in addition to impact zones in an attempt to move toward a more quantitative treatment of the problem. An automated analysis function will also be implemented to analyze Natech risk for facilities available in the RAPID-N database immediately after the occurrence of a major natural event, so that competent authorities, first responders and other interested parties can be alerted quickly to ensure fast protective action if required.

8.2 PANR

A qualitative methodology for the preliminary assessment of Natech risk (PANR) in urban areas was proposed by Cruz and Okada (2008). The PANR methodology involves identifying, quantifying, and analyzing the risk posed by the presence of hazardous materials in a territory subject to natural hazards to local exposed elements in a community. PANR defines risk as a function of the hazard (magnitude and probability) and the vulnerability of the elements exposed (consequences, their severity, and probability) to the hazards (both the natural and secondary hazards from any chemical accidents and their domino effects).

$$\text{Risk} = \text{Hazard} * \text{Vulnerability} \qquad (8.1)$$

The PANR methodology involves several steps including data collection and inventory development, hazard identification and vulnerability analysis, and estimation of a Natech risk index for each storage tank containing hazmats in a territory. Once the Natech risk-index values have been estimated, it is possible to identify those areas with high Natech risk (Natech hotspots). Using the previous expression for risk, the Natech risk index (NRI_i) for each hazmat containing storage tank i in a territory for a given natural disaster scenario is defined as follows:

$$[\text{NRI}_i] = [\text{HRL}_i] * [D_i + \text{Area-sc}_i + C_i] \qquad (8.2)$$

where HRL is a score that accounts for the hazmat-release likelihood, given the natural-hazard event; D is a score that accounts for the effects of potential domino

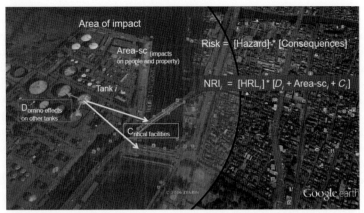

FIGURE 8.4 Graphic Representation for the Natech Risk Index (NRI$_i$) for Tank i (Base Map ©2016 Google, Zenrin)

chemical accidents; Area-sc is a score that measures the potential consequences of the hazmat release from each tank i on the population in the directly affected area given the natural-disaster scenario; and C is a score that measures the potential consequences of the hazmat release from each tank i on essential facilities located within the directly impacted area that are critical for the safety and well-being of the community and the environment given the natural-disaster scenario. Fig. 8.4 shows a schematic of these relationships.

The PANR methodology was developed as a diagnostic tool for local communities and is intended to promote the participation in the assessment process of local government officials and first responders in consultation with community members, industrial-facility operators, and experts. The PANR methodology was tested among representatives of European Union Member States during an international workshop at the Joint Research Centre of the European Commission in 2007 (Cruz and Krausmann, 2008) to identify minimum skills and knowledge needed to carry out the assessment.

Improvements to the methodology were proposed by Cruz and Suda (2015) by introducing relationships to estimate D, Area-sc, and C, as well as implementing PANR into the Joint Research Centre's RAPID-N tool (Section 8.1) for the estimation of the HRL values. The authors applied the methodology to an industrial park and neighboring residential areas in Kobe, Japan, and compared the results obtained with estimates of NRIs for past Natech accidents in Japan. Data for the study were collected through interviews, survey questionnaires, site visits, review of government reports, city plans, chemical-accident rules and regulations, as well as past studies and peer-reviewed literature. In the future, the authors plan to carry out the assessment process involving local community members, industry representatives, and local government authorities.

8.3 TRAS 310 AND TRAS 320

The German Major Accidents Ordinance, which transposes the main part of the EU Seveso Directive into national law, requires in its Art. 3 that the operator of an establishment, where large amounts of certain hazardous substances are used or may be present, has to take the required precautions to prevent and in case of a failure take measures to limit the possible consequences of major accidents like the release of hazardous substances, fires, and explosions. These precautions and measures have to reduce the risks of major accidents according to the state of the art in safety. In this context, the operator has to consider natural hazards for the determination of the necessary precautions and measures.

The German Federal Ministry for the Environment is allowed to issue *Technical Rules for Installation Safety (TRAS)* which concretize these obligations of operators. Related to Natech risks this was done with:

1. a TRAS for hazards triggered by floods and precipitation (TRAS 310) and
2. a TRAS for hazards triggered by wind, snow loads, and ice loads (TRAS 320).

Both were drafted by the German Commission on Process Safety [Kommission für Anlagensicherheit (KAS)], appointed according to Art. 51a of the Federal Immission Control Act [Bundes-Immissionsschutzgesetz (BImSchG)].

Both TRAS are valid for establishments that fall within the scope of the Major Accidents Ordinance. However, it is recommended that both TRAS should also be applied to all other installations not subject of the Major Accidents Ordinance but that require licensing according to the BImSchG and if a risk of an accident involving hazardous substances cannot be excluded. Therefore the term "sites" is used if an obligation of a TRAS is relevant for establishments and should be applied to theses installations as well.

Both TRAS follow a similar approach which is flexible and based on methodologies which are applied already to operational hazards. In addition, both TRAS define probabilities or intensities of the addressed natural hazards to be taken into consideration in the design and operation of installations. Moreover, in the case of TRAS 310 (precipitation and floods), the expected effects of climate change in Germany on these natural hazards are taken into account by considering a "climate-change factor." In the case of TRAS 320 (wind, snow loads, and ice loads), the relevant expected effects of climate change in Germany were evaluated and it was decided that according to the current state of knowledge a "climate-change factor" is not appropriate. Therefore, the TRAS includes other regulations on the adaptation of protection aims and installations to climate change.

8.3.1 TRAS 310 "Precautions and Measures Against the Hazard Sources Precipitation and Flooding"

8.3.1.1 *Scope of Application*

TRAS 310 (TRAS 310, 2012a,b; Köppke et al., 2012) is valid for sources of hazards to sites (later called "hazard sources") which result from

1. floods caused by riverine and flash floods or storm surge, including the failure of flood defenses;
2. drainage flooding, e.g., caused by heavy precipitation or sewer backup; and
3. rising groundwater.

From the hazards that may be linked with precipitation and floods, only flotsam is addressed in TRAS 310. Nevertheless, due to the obligations according to Art. 3 of the German Major Accidents Ordinance, operators of establishments have to consider other related hazards like hail, landslides, etc.
The following sections on TRAS 310 include translations of the text of the TRAS.

8.3.1.2 Methodological Approach of TRAS 310
The methodological procedure of TRAS 310 is illustrated in Fig. 8.5. The operator's obligations may be fulfilled with regard to hazard sources by following these four actions:

1. Hazard source analysis to investigate the impact on the site for each hazard source as a single hazard or in combination with other natural hazards.
2. Analysis of hazards and threats to examine their impacts on each safety-relevant part of an establishment or installation.
3. Elaboration of a protection concept against major accidents (here Natech accidents).
4. Examination of "major accidents despite precautions," which leads in particular to the specification of measures to mitigate the effects of (nevertheless occurring) major accidents.

The only new part of this approach is the "hazard source analysis." The other steps are in line with the approaches used traditionally in the preparation of safety reports.

8.3.1.3 Hazard Source Analysis
The starting point of the methodology is a hazard source analysis in which the possible sources of hazards to the sites are determined. First, a simplified hazard source analysis identifies those natural events which cannot "reasonably be excluded" at the site locations in qualitative terms. Second, in a detailed hazard source analysis, further information must be collected to determine probabilities and intensities of the possible hazard sources.

8.3.1.3.1 SIMPLIFIED HAZARD SOURCE ANALYSIS
First it is to be determined which hazard sources occurred in the past or may occur at the site location and which can reasonably be excluded. Simple criteria can be applied to make this determination. Natural-hazard maps, especially flood maps, play an important role in this regard. However, information in old maps must be treated with caution, and it needs to be ascertained that

1. the date of map preparation is known to check the reliability of the data (orography, currency, level of detail) and

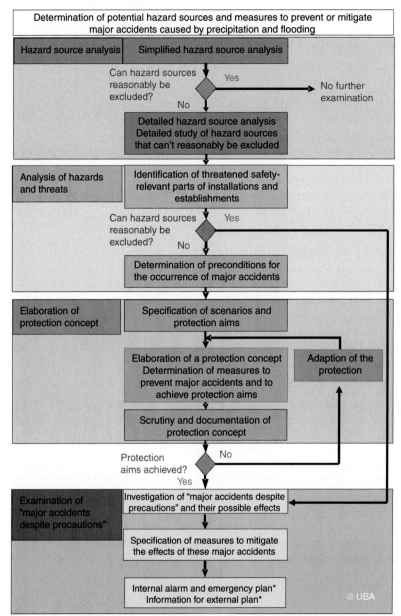

FIGURE 8.5 Flowchart for the Consideration of Natural Hazards in Safety Management

*If required pursuant to Art. 10 of the Major Accidents Ordinance only.

Table 8.1 Criteria for Selected Natural-Hazard Sources

Hazard Source	Criterion	Necessity and Extent of Hazard Source Analysis	
River or coastal high water flooding combined with flow, dynamic pressure, flotsam, and ice run	Designated flood plain or area mapped on (flood) hazard or risk maps under Art. 74 of the Federal Water Act	On the designated flood plain or within the mapped (flood) risk area	Detailed hazard source analysis required
		Mapped, but outside (flood) risk areas	No further examination required
Rising groundwater	Underground parts of installations where hazardous substances are present (tanks, pipes)	Releases due to buoyancy possible	Detailed hazard source analysis required
		Releases due to buoyancy not possible	No further examination required

2. all relevant kinds of hazards are considered (often flash floods and floods from sewers are not included).

Selected criteria are given in Table 8.1 for riverine and coastal flooding, potentially combined with flow, flotsam, and ice run. No simple, general criterion can be cited that makes it reasonably possible to exclude the hazard source "flooding" triggered by precipitation outside mapped (flood) risk areas.

8.3.1.3.2 DETAILED HAZARD SOURCE ANALYSIS

Where hazard sources cannot reasonably be excluded, a detailed hazard source analysis is required. The following trigger events are to be assumed for the detailed hazard source analysis:

1. As foundation for precautions to prevent major accidents (first obligation according to Art. 3 of the Major Accidents Ordinance):
   a. events with medium probability (recurrence interval at least 100 years according to Art. 74 of the Federal Water Act) and requirements due to climate change according to Annex I of TRAS 310, or
   b. events with a higher recurrence interval (>100 years) in those cases where the site is located next to water bodies (e.g., rivers or coast) and if public defense structures (e.g., dikes) are designed for a higher recurrence interval (e.g., 200 years).
2. Generally, it must be assumed that water penetrates into the site with at least an intensity expected for a 100-year event to determine the measures required to limit the consequences of major accidents (second obligation according to Art. 3 of the Major Accidents Ordinance).

With regard to the basis for measures to prevent major accidents, operators have to consider that the standards for the dimensioning of public flood defenses may be based on events that occur more rarely than 100-year events. For sites behind these defenses, operators usually do not have to take any own precautions to prevent major accidents, if the failure of the defenses can be reasonably excluded as a hazard source (Art. 3 of the Major Accidents Ordinance). However, if a site is directly situated next to water, the standards for the dimensioning of public flood defenses up- and downstream of the site need to be applied to the flood defense of this site. Otherwise there would be a gap in the flood defenses and when a flood occurs, the water could flow through the site into the hinterland.

If the flood defenses have not been dimensioned, constructed, and maintained in accordance with the relevant technical rule for dikes, their failure cannot be excluded. In this case (i.e., for old dikes) the operator has to either take own flood defense precautions or contribute to remedial works of the public flood defenses.

The detailed hazard source analysis involves the following steps:

1. Determination of the potential inflow and runoff routes with direction of flow.
2. Determination of possible water levels dependent on the intensity of the event.
3. Quantification of possible flow speeds (to estimate the effects of dynamic pressure and flotsam).
4. Estimation of the threat from flotsam or ice run.
5. Estimation of the threat from erosion (undermining of buildings and parts of installations).
6. Estimation of the threat from the flotation of installations and parts of installations.

For Step 2, an inflow–outflow calculation is useful. Flooding of a site is only possible if the inflow of water is significantly greater than the runoff of water. For this reason, potential inflow routes and rates must be compared by the operator with the runoff routes and rates. A potential inflow of water may be caused by (Fig. 8.6)

1. extreme precipitation,
2. water backing up from the sewer system (onsite/offsite),
3. surface water (lateral inflow due to terrain formation, e.g., at locations in depressions),
4. lateral inflow due to high-water flooding or the failure of flood defenses (levee, gates), or
5. groundwater or return seepage.

On the other hand, relevant runoff routes may be (Fig. 8.7)

1. surface runoff (due to terrain formation, harmless diversion of excess water along roads when extreme events occur),
2. seepage to aquifers,
3. sewer systems (onsite/offsite), and
4. flood pumping stations (along waters).

In this context it should be borne in mind that the probabilities of failure used in the design of sewer systems are much higher than those for the design of flood defenses. This means that in case of heavy precipitation events, sites can have a significant inundation risk due to sewer failure even if they are located far away from any surface water.

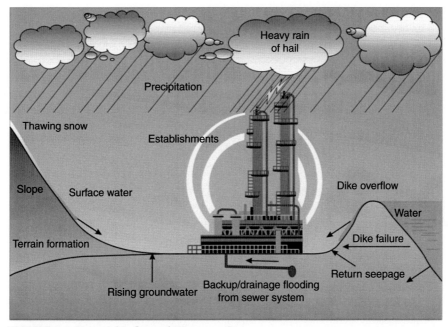

FIGURE 8.6 Potential Inflows of Water at a Site

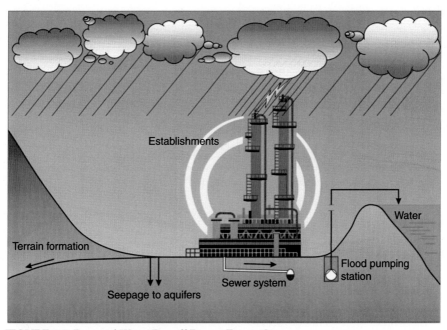

FIGURE 8.7 Potential Water Runoff Routes From a Site

If flooding cannot be excluded, digital terrain models and computer programs for hydrological and hydraulic simulation should be applied. For simple cases, a 1D representation can be reasonable, for larger sites or in case of unclear flow directions, a 2D simulation is useful.

Neoformation of groundwater may be caused by sustained rain or by flooding. It occurs with time delay and leads to a rise of the groundwater level. This increases the buoyancy of underground tanks and pipes and may cause a threat for underground parts of an installation. In order to assess the threat from a rise in groundwater, information on the groundwater level is required and calculation models should be applied.

8.3.1.3.3 CONSIDERATION OF CLIMATE CHANGE

The foreseeable consequences of climate change should be taken into consideration in the course of a hazard source analysis, even if there are uncertainties.

With the global temperature rising as a consequence of climate change, the atmosphere's capacity to absorb water vapor will increase proportionately. This gives reason to believe that the intensity and frequency of heavy precipitation events will increase in line with the rise in temperature. Therefore, the preconditions for flooding should be adapted to more frequent and intense heavy precipitation. According to the emission scenarios discussed by the IPCC (2007), it is to be assumed that the precipitation volumes in winter in Germany could be 0–15% higher over the period 2021–50 than during the control period 1961–90. Over the period 2071–2100, they could be 0–40% higher, while the regional volumes of precipitation may vary widely.

As it has still not been possible to determine scientifically a climate-change factor for each region, a standard climate-change factor of 1.2 should be applied as a matter of precaution, unless the consequences of climate change have already been taken into consideration by the competent authorities pursuant to Articles 72–81 of the Federal Water Act in their (flood) hazard maps, or the competent water authority has previously determined possible changes due to climate change in runoff models of floods (for enforcement see Section 8.3.1.12).

8.3.1.4 Determination of Threatened Safety-Relevant Parts of Establishments and Installations

If the impacts by precipitation and floods which cannot reasonably be excluded are known, it is to be determined which safety-relevant parts of installations and establishments could be affected. The safety-relevant parts of establishments and installations of this kind are as follows:

1. Installations and parts of installations where hazardous substances are present.
2. Installations and parts of installations with safety-relevant functions.

The threatened installations can be easily identified by comparing their elevation with the expected water level of a flood which occurs once in 100 or more years.

8.3.1.5 Determination of Possible Causes of Major Accidents

For the identification of the possible causes of major accidents, it is to be determined for each threatened part of establishments and installations how the hazard source that would become active could affect the safety-relevant parts of the installations and establishments threatened in the specific case. The following approach is proposed:

1. Determination of the effects on threatened parts of installations where hazardous substances are present.
2. Determination of the effects on threatened parts of installations with safety-relevant functions (within installations).
3. Determination of the effects on threatened installations where hazardous substances are present.
4. Determination of the effects on threatened installations with safety-relevant functions inside and outside the site.
5. Determination of the effects on the whole establishment or installation.

In the last step, the consequences of the simultaneous effects of hazard sources on all parts of the establishments and installations on the site, and the interactions between them (effects on one installation/part of an installation trigger a major accident in another installation/part of the same installation) are to be examined.

8.3.1.6 Specification of Scenarios and Protection Aims

Having identified the possible hazard sources (TRAS Section 7 "Detailed Hazard Source Analysis") and the possible hazards or threats which they could trigger (TRAS Section 9 "Determination of Preconditions for the Occurrence of Major Accidents"), scenarios to cover these hazard sources are to be determined and examined in detail. It has to be proved that the effectiveness of precautionary measures is consistent with the state of the art of safety technology according to Art. 3 of the Major Accidents Ordinance.

In the investigation of the scenarios, the primary protection aims are the protection of people, the environment, and property (Art. 5 of the BImSchG and Art. 3 of the Major Accidents Ordinance). Precautions have to consider these general protection aims and are to be specified in concrete terms in relation to the hazard sources and the associated scenarios.

The basis for the specification in concrete terms are the results of the hazard source analysis, from which information is derived about the intensity of a hazard source as a function of its probability of occurrence. If the damage triggered by hazardous sources of different intensities is known, the risks can be determined. These risks must be reduced to an accepted level by defining the general protection aims in concrete terms.

An at least 100-year event should be taken as the basis for the specification of the protection aims. They should be based on rarer events if they are used for the design

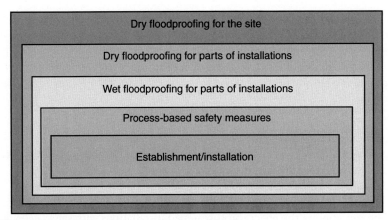

FIGURE 8.8 Safety Precautions and Measures (Flooding)

of public flood defenses with regard to sites that are directly located next to water bodies. The consequences of climate change for the various hazard sources are to be taken into consideration additionally (see Annex I of TRAS 310).

8.3.1.7 Elaboration of Protection Concepts for Scenarios

Protection concepts, as a part of the operator's safety management, are to be developed for those hazard sources that cannot reasonably be excluded. The hazards or threats must be identified, and the scenarios and protection aims be determined.

Every protection concept should include various safety precautions and measures (lines of defense) (Fig. 8.8). For existing establishments and installations, and those to be constructed, different kinds of precautions and measures can make sense.

8.3.1.8 Review of Protection Concepts

The protection concept developed must be reviewed with a focus on the achievement of the protection aims. This process has to verify the probabilities of occurrence, the intensities of the natural-hazard sources, and the probabilities of failure of the precautions and measures chosen to reduce the risk.

The review serves as verification of the obligations of the operator according to the Major Accidents Ordinance and the BImSchG. If the chosen precautions and measures are considered to be insufficient, the protection concept in question is to be revised in order to incorporate further precautions and measures to provide for major accidents.

8.3.1.9 Determination of Accident Scenarios

According to the Major Accidents Ordinance the operator has to elaborate accident scenarios. These scenarios are drawn up to determine the following:

1. The measures required to mitigate the effects of major accidents that can reasonably be excluded pursuant to Art. 3 Para. 3 of the Major Accidents Ordinance (major accidents despite precautions).
2. The information required for the elaboration of internal alarm and emergency plans pursuant to Art. 10 of the Major Accidents Ordinance.

3. The information required for the drafting of external alarm and emergency plans pursuant to Art. 9 of the Major Accidents Ordinance.

Items 2 and 3 are only required if establishments are subject to the "upper tier" requirements of the Major Accidents Ordinance.

Concerning hazards caused by precipitation and floods, attention is to be paid to the following points when these scenarios are drafted:

1. Parts of installations located at higher elevations may not have to be taken into consideration as a consequence of the exclusion of exceptional events.
2. Natural-hazard sources, for example, flooding, may affect several parts of an installation simultaneously and cause disturbances.
3. As a consequence, more than the largest mass contained in a part of an installation may be released under certain circumstances (e.g., leakage of several tanks).
4. Apart from the dispersion of substances in the atmosphere, when events caused by flooding and precipitation occur, dispersion in water is to be assumed.
5. It is to be assumed that the availability of measures to mitigate consequences will be limited in case of impacts by a natural hazard (e.g., limited availability of access routes, etc.).
6. In addition, it is to be assumed that the availability of external personnel will be limited.
7. Moreover, the extent to which an impact may trigger another hazard at another installation or another part of the same installation is to be evaluated.

8.3.1.10 Specification of Measures to Mitigate the Effects of Major Accidents

According to Art. 3 Para. 3 of the Major Accidents Ordinance, the operator has to take precautionary measures to keep the effects of major accidents as small as possible. Whether and to what extent the external natural-hazard sources examined in the TRAS 310 require measures of any kind to prevent the dispersion of contaminants must be evaluated systematically in the individual case.

8.3.1.11 Planning for Emergencies

The operator with "extended obligations" according to the Major Accidents Ordinance ("upper tier establishments" subject to the Seveso Directive) has to set up and update if required an internal alarm and emergency plan, and has to supply information required for external alarm and emergency plans. The results of the previously mentioned steps can require updates of both.

8.3.1.12 Design Criteria for the Consideration of Climate Change

Annex I of the TRAS 310 includes the following design criteria for sites which have to take the expected effects of climate change in Germany into consideration. The principles for the purpose of adaptation to climate change are as follows:

1. A climate-adaptation factor of 1.2 is applied to the triggering natural-hazard intensities (for a probability of 1 in 100 years) to be estimated for 2010 to take into consideration possible changes in the period up to 2050.

Table 8.2 Requirements for Adaptation to Climate Change

Hazard Source	Intensity to Be Estimated as of 2010	Intensity to Be Estimated for 2050
River flooding	Flood runoff (m³/s)	1.2 × flood runoff (m³/s)
Flash flood events	Flood runoff (m³/s)	1.2 × flood runoff (m³/s)
Storm surge events	Nominal height of levees, etc. pursuant to designation	May subsequently be raised by up to 1 m[a]
Heavy precipitation	Peak heavy precipitation[b] for $t = 100$ a	1.2 × peak heavy precipitation for $t = 100$ a
Rising groundwater	Surface of terrain	Surface of terrain (climate change adaptation factor not relevant)

[a]*General coastal defense plans, for example, the measures taken by the Lower Saxony Water, Coastal Defense and Nature Conservation Agency, http://www.nlwkn.niedersachsen.de*
[b]*http://www.dwd.de/kostra*

2. New installations that will be designed for the period up to 2050 or after should comply with the requirements.
3. The climate-adaptation factor does not have to be taken into consideration if the intention is to operate a planned new installation not until 2050.
4. As of 2050, the climate-adaptation factor is to be considered in the layout of all installations (i.e., including the existing ones).
5. A detailed hazard source analysis may provide reasons for a variation of the 1.2 factor in individual cases. This is possible in particular if the consequences of climate change are already taken into consideration in (flood) hazard maps or the competent water authority has previously ascertained the possible change in the runoff flooding due to climate change.
6. Should other developments in what is known about climate come to light in the period up to 2050, they will be taken into consideration when this Technical Rule on Installation Safety is revised.

The need for adaptation to climate change is taken into consideration in the requirements presented in Table 8.2.

8.3.2 TRAS 320 "Precautions and Measures Against the Hazard Sources Wind, Snow Loads and Ice Loads"

TRAS 320 (2015a), which is relevant for the natural hazards wind, snow loads, and ice loads, was issued in Jul. 2015. It is based on the methodology already used in TRAS 310. The details of TRAS 320 are not presented here but the interested reader is referred to a complete translation in English (TRAS 320, 2015b). In addition, recommendations and explanations to TRAS 320 are available in German (Krätzig et al., 2015).

8.3.3 Summary

Both TRAS have to be seen as rules to implement the requirements of the German Major Accidents Ordinance considering approaches applied already in its enforcement in Germany. They both propose the same methodology which is based in its Steps 2–4 on "traditional" hazard analysis and risk management in Germany. The new introduced instrument is the first step, the "Hazards Source Analysis." This step requires first a qualitative and then a quantitative analysis of probabilities and intensities of natural hazards at site locations. Natural-hazard maps are very helpful in this step but may not cover all relevant natural hazards. The challenge is to find sufficient historical data for the evaluation of these other relevant natural hazards.

TRAS 310 may be one of the first technical rules considering the expected consequences of climate change. This was possible due to enormous work carried out in Germany, especially on projections of climate change at the regional level. TRAS 320 includes no "climate-change factor" in its requirements but the obligation to consider exceptional snow loads in all parts of Germany. This reflects the discussions on climate-change effects in winter.

The main challenge of both TRAS is the cooperation of experts for industrial safety science with other disciplines. Therefore, the elaboration of the TRAS requirements had to consider the risk-management approaches of these other disciplines, as well. As flood and precipitation risk management, as well as civil engineering (like DIN EN 1990 and 1991) are based on semiprobabilistic approaches, both TRAS had to do the same. In addition, both TRAS had to address hazards that are usually not considered by these other disciplines but are relevant for Natech risk management (e.g., hazards by flotsam or airborne projectiles). For these hazards, deterministic requirements were added.

In the final state of their preparation, both TRAS were "tested," that is, implemented at establishments or installations. TRAS 310 was implemented at a chemical establishment threatened by a small river; TRAS 320 was implemented related to wind impacts for a distillation column in a coastal area in the north of Germany, and for a tank close to the coast of the Baltic Sea in Germany related to snow load. All cooperating operators regarded the methodology of the TRAS as a useful approach. In the case of the "TRAS 310 test site," the site was designed in such a way that a flood could damage parts of the site but would not be able to cause a major accident. Two years after the test exercise, the site was exposed to severe flooding and the consequences were as assumed.

The application of TRAS 320 related to wind hazards showed that the increase of the requirements of the DIN standards used in civil engineering (introduction of wind zones in 2005 with an increased design speed at that site) was more relevant than the additional requirements of TRAS 320. The application of TRAS 320 and its requirements related to snow, on the other hand, showed that availability of data on the design and construction of installations is very relevant. The absence of information on the static design and gaps in the documentation of the construction of installations may be more costly than the implementation of the TRAS 320.

8.4 OTHER METHODOLOGIES

Methods for seismic risk assessment at industrial plants were first proposed in the 1980s. Reitherman (1982), for example, offered some suggestions on engineering approaches for the prevention of earthquake-triggered spills after studying releases from a number of smaller earthquakes during the period of 1964–80. Kiremidjian et al. (1985) developed a general methodology for seismic risk analysis at major industrial facilities, focusing on methodologies for estimating the damage to structures and equipment. Werner et al. (1989) identified potential hazardous-materials releases that could occur in Silicon Valley facilities and suggested a methodology for risk mitigation. Tierney and Eguchi (1989) described a methodology for estimating the risk of postearthquake hazardous-materials releases of anhydrous ammonia and chlorine in the Greater Los Angeles Area. The pilot application of the methodology was discussed in detail by Seligson et al. (1996).

More recently, Busini et al. (2011) proposed a qualitative screening tool for seismic Natech risk using the analytical hierarchy process as multicriteria decision model for evaluating suitable qualitative key hazard indicators. This methodology facilitates the identification of situations in which a more complex and costly quantitative risk assessment is called for.

In the state of California, the Administering Agency's Subcommittee, Region I Local Emergency Planning Committee (LEPC) updated the California Accidental Release Prevention (CalARP) Program Seismic Assessment Guidance in 2013 (CalARP, 2013). The guidance specifically recommends that installations perform the seismic assessment on

1. covered processes (those where regulated substances are stored, processed, or otherwise handled) as defined by CalARP Program regulations;
2. adjacent facilities whose structural failure or excessive displacement could result in the significant release of regulated substances; and
3. onsite utility systems and emergency systems which would be required to operate following an earthquake for emergency reaction or to maintain the facility in a safe condition, (e.g., emergency power, leak detectors, pressure relief valves, battery racks, release treatment systems including scrubbers or water diffusers, firewater pumps and their fuel tanks, cooling water, room ventilation, etc.).

Most importantly, the guidance recommends that in order to reduce the risk of accidental releases, individual equipment items, structures, and systems (e.g., power, water, etc.) may need to achieve varied performance criteria. The guidance includes the following four main criteria: (1) maintain structural integrity, (2) maintain position, (3) maintain containment of material, and (4) function immediately following an earthquake. In addition, the updated guidance specifically addresses the assessment of ground shaking, including local site amplification effects, fault rupture, liquefaction and lateral spreading, seismic settlement, landslides, and tsunamis and seiches. In addition, it provides guidance on prevention and mitigation measures to reduce the risk of hazmat releases (CalARP, 2013).

Similarly, the Institute for Disaster Mitigation of Industrial Complexes at Waseda University in Japan has published comprehensive Guidelines for Earthquake Risk Management at Industrial Parks located in coastal areas (IDMC, 2016) taking into account the lessons learned from the Great East Japan earthquake and tsunami.

Ayrault and Bolvin (2004) developed a methodology for the integration of flood hazards in the risk-reduction process at industrial facilities. It recommends carrying out a risk analysis for each type of equipment that could cause a major accident following flood damage. The methodology also emphasizes the identification of safety barriers based on the predicted accident scenarios. Also El Hajj et al. (2015) address flood impacts at hazardous installations. They developed a qualitative risk-analysis methodology that includes the development of generic reference bow-ties with a focus on accident scenarios triggered by floods. The methodology was validated through application of the accident scenarios in the surface treatment sector.

References

Ayrault, N., Bolvin, C., 2004. Analyse des risques et prévention des accidents majeurs—Rapport partiel d'opération f: Guide pour la prise en compte du risque inundation, INERIS Report DRA-2004-Ava-P46054 (in French).

Busini, V., Marzo, E., Callioni, A., Rota, R., 2011. Definition of a short-cut methodology for assessing earthquake-related Na-Tech risk. J. Hazard. Mater. 192, 329.

CalARP, 2013. Guidance for California Accidental Release Prevention (CalARP) Program Seismic Assessments, CalARP Program Seismic Guidance Committee, December 2013.

Cruz, A.M., Okada, N., 2008. Methodology for preliminary assessment of Natech risk in urban areas. Nat. Hazards 46, 199.

Cruz, A.M., Krausmann, E., 2008. Results of the workshop: Assessing and managing Natechs (natural-hazard triggered technological accidents), JRC Scientific and Technical Report EUR 23288 EN, European Communities.

Cruz, A.M., Suda, H., 2015. Assessment of Natech risk for evacuation planning in areas subject to natural hazards. In: Proceedings of Seventh European Meeting on Chemical Industry and Environment, Tarragona, Spain, June 10–12.

El Hajj, C., Piatyszek, E., Tardy, A., Laforest, V., 2015. Development of generic bow-tie diagrams of accidental scenarios triggered by flooding of industrial facilities (Natech). J. Loss Prevent. Process Ind. 36, 72.

FEMA, 2003. Multi-Hazard Loss Estimation Methodology, Earthquake Model, HAZUS®MH MR4, Technical Manual. US Federal Emergency Management Agency, Washington, DC.

Girgin, S., 2012. RAPID-N: Rapid Natech risk assessment tool—User manual version 1.0, JRC Technical Report EUR 25164 EN, European Union.

Girgin, S., Krausmann, E., 2013. RAPID-N: rapid Natech risk assessment and mapping framework. J. Loss Prevent. Process Ind. 26, 949.

IDMC, 2016. 臨海部コンビナート施設の地震リスクマネジメントガイドラインを発行しました。(Guidelines for Earthquake Risk Management at Industrial Parks located in Coastal Areas). Institute for Disaster Mitigation of Industrial Complexes, Waseda University, Tokyo, April (in Japanese).

IPCC, 2007. Climate Change 2007: Impacts, Adaptation and Vulnerability, Fourth Assessment Report. Cambridge University Press.

Kiremidjian, A., Ortiz, K., Nielsen, R., Safavi, B., 1985. Seismic Risk to Industrial Facilities, Report No. 72. The John A. Blume Earthquake Engineering Center, Stanford University.

Köppke, K.E., Sterger, O., Stock, M., 2012. Hinweise und Erläuterungen zur TRAS 310 Vorkehrungen und Maßnahmen wegen der Gefahrenquellen Niederschläge und Hochwasser. Umweltbundesamt, Dessau-Roßlau, April 2012, http://www.kas-bmu.de/publikationen/tras/370849300_TRAS310_Endbericht_1.pdf (in German).

Krätzig, W.B., Andres, M., Niemann, H.-J., Köppke, K.E., Stock, M., 2015. Hinweise und Erläuterungen zur TRAS 320 - Vorkehrungen und Maßnahmen wegen der Gefahrenquellen Wind, Schnee- und Eislasten, Umweltbundesamt, Dessau-Roßlau, October 2015, http://www.kas-bmu.de/publikationen/tras/TRAS320_Hinweise_Erlaeuterungen.pdf (in German).

Krausmann, E., Baranzini, D., 2012. Natech risk reduction in the European Union. J. Risk Res. 15, 1027.

Reitherman, R.K., 1982. Earthquake-caused hazardous materials releases. In: Proceedings of Hazardous Materials Spills Conference, Milwaukee, WI, April 19–22.

Seligson, H.A., Eguchi, R.T., Tierney, K.J., Richmond, K., 1996. Chemical Hazards, Mitigation and Preparedness in Areas of High Seismic Risk: A Methodology for Estimating the Risk of Post-Earthquake Hazardous Materials Releases, Technical Report NCEER-96-0013. National Center for Earthquake Engineering Research, State University of New York at Buffalo.

Tierney, K.J., Eguchi, R.T., 1989. A methodology for estimating the risk of post-earthquake hazardous materials releases. In: Proceedings of HAZMACON 89, Santa Clara, California, April 18–20, 1989.

TRAS 310, 2012a. Technische Regel für Anlagensicherheit: Vorkehrungen und Maßnahmen wegen der Gefahrenquellen Niederschläge und Hochwasser, Bundesanzeiger, February 24, 2012, Supplement 32a, non-official version: http://www.kas-bmu.de/publikationen/tras/TRAS310_nicht_offizielle_Version.pdf (in German).

TRAS 310, 2012b. Technical Rule on Installation Safety: Precautions and Measures Due to Precipitation and Floods, May 2012, non-official short version: http://www.kas-bmu.de/publikationen/tras/TRAS_310_GB_shortversion.pdf

TRAS 320, 2015a. Technische Regel für Anlagensicherheit: Vorkehrungen und Maßnahmen wegen der Gefahrenquellen Wind sowie Schnee-und Eislasten, Bundesanzeiger, July 16, 2015, non-official version: http://www.kas-bmu.de/publikationen/tras/tras_320.pdf (in German).

TRAS 320, 2015b. Precautions and Measures Against the Hazard Sources Wind, Snow Loads and Ice Loads, June 15, 2015, translation into English: http://www.kas-bmu.de/publikationen/tras/TRAS_320_en.pdf (in English).

US EPA, 1999. Risk Management Program Guidance for Offsite Consequence Analysis, US Environmental Protection Agency, Chemical Emergency Preparedness and Prevention Office, USA.

Werner, S.D., Boutwell, S.H., Varner, T.R., 1989. Identification and mitigation of potential earthquake-induced hazardous material incidents in Silicon Valley facilities. In: Proceedings of HAZMACON 89, Santa Clara, California, April 18–20.

Quantitative Methods for Natech Risk Assessment

V. Cozzani, E. Salzano
Department of Civil, Chemical, Environmental, and Materials Engineering,
University of Bologna, Bologna, Italy

Quantitative risk assessment is a powerful but complex and time-consuming task, which requires a significant amount of information and sophisticated models for the analysis of a very high number of scenarios even for rather simple plant layouts. The availability of software tools that support the risk analyst is therefore crucial. In this chapter, two tools that support the quantitative analysis of Natech risk are presented. The risk figures resulting from the application of these tools can then be used for comparison with quantitative risk-acceptability criteria.

9.1 ARIPAR

9.1.1 Framework of the ARIPAR-GIS Natech Module

The Natech module of the ARIPAR-GIS software was developed to implement the specific procedure for the quantitative analysis of Natech events developed by Antonioni et al. (2009). The procedure is summarized in Fig. 9.1 and is actually a customization of the general framework presented in Fig. 7.3. The specific assumptions and steps introduced to implement the general methodology discussed in Chapter 7 are briefly summarized later in the chapter.

The starting point in the quantitative assessment of Natech scenarios is the characterization of the frequency and severity of the natural event by a sufficiently simple approach, which is suitable for use in a risk-assessment framework (Steps 1–3 in Fig. 9.1A). Usually, in this step a limited number of "reference events" is identified, each having a given intensity and an expected frequency or return time. A set of impact vectors may thus be defined, the elements of the vectors being the intensity of the natural events characterized by one or more intensity parameters selected to describe the natural event. It should be noted that this step in no way is intended to provide a characterization of the natural hazard at the site, nor to provide data for a detailed analysis of the damage to structures, but only to obtain the input data necessary for use in simplified equipment damage models.

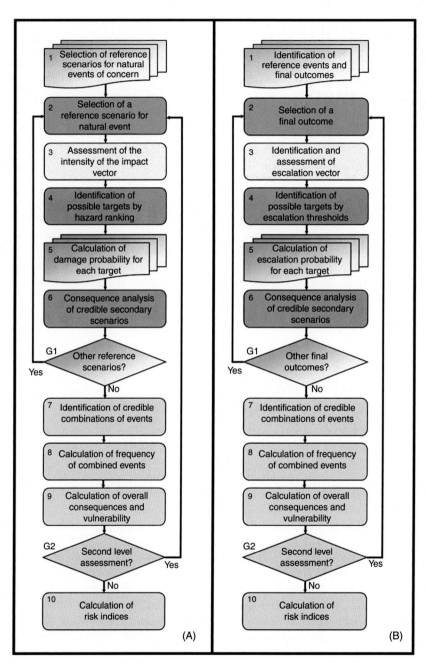

FIGURE 9.1

Flowchart of the Framework Proposed for Quantitative Risk Assessment of (A) Natech Scenarios and (B) Domino Effect

Adapted from Cozzani et al. (2014).

Simplified hazard-ranking criteria based on inventory and physical state of hazardous substances may be used to identify critical equipment items that should be included in the analysis (Step 4 in Fig. 9.1A) (Antonioni et al., 2009). The application of vulnerability models is then needed to assess the equipment damage probability (Step 5). These equipment vulnerability models are further discussed in Section 9.1.5. Consequence assessment of the single scenarios triggered by the natural event (Step 6) can be carried out by using conventional models, although a limited number of Natech-specific final outcomes may arise (Cozzani et al., 2010; Renni et al., 2010).

The final steps of the procedure (Steps 7–10) are aimed at risk recomposition. These steps require a dedicated approach for the identification of possible multiple and simultaneous accident scenarios and the calculation of their frequencies and consequences. This procedure, which is summarized in Table 9.1, was originally developed within the framework of the risk analysis of domino accidents presented in Fig. 9.1B (Cozzani et al., 2005, 2006). Actually, as shown in the figure, several steps needed in the quantitative analysis of risk due to either domino effects (cf. Section 7.3.2.2) or the impact of natural events on process equipment are similar, and similar mathematical procedures can be applied in the assessment process. The procedure used for Steps 7–10 is described in detail in Antonioni et al. (2007), Reniers and Cozzani (2013), and Cozzani et al. (2014).

9.1.2 The ARIPAR-GIS Software

The ARIPAR-GIS software was developed in the framework of the ARIPAR project (Egidi et al., 1995), which was one of the first applications of Quantitative Area Risk Analysis techniques in the evaluation of all hazards in an extended industrial area. The ARIPAR-GIS software allows the calculation of individual and societal risk originating from multiple risk sources due to both fixed installations and hazardous-materials transport systems. The software is supported by a geographical information system (GIS) platform that allows positioning of the different risk sources and producing risk maps as a result of the assessment. Spadoni et al. (2000, 2003) provide a detailed description of the software.

9.1.3 The Natech Package of the ARIPAR-GIS Software

A specific software package was developed and added to the ARIPAR-GIS software in order to allow the quantitative analysis of risk due to Natech events. The implemented procedure allows the automatic identification of all the possible overall scenarios that may be generated by the impact of a natural event (earthquake or flood) on a hazardous site of interest. A simplified layout needs to be implemented in a GIS environment. The procedure automatically generates all the overall events generated by equipment damage, calculates the expected overall frequencies and vulnerability maps, and performs the quantitative analysis of the risk in the area of interest.

Table 9.1 Summary of Steps 7–10 for the Identification of Credible Combinations of Events and of the Resulting Frequencies and Consequence Evaluation, Taking Into Account Multiple Simultaneous Failures

Item	Definition	Value/Equation
Input Parameters		
n	Total number of target equipment	—
k	Number of target equipment simultaneously damaged by a Natech scenario	—
N_k	Number of Natech-induced scenarios involving k different final outcomes	$N_k = \begin{pmatrix} k \\ n \end{pmatrix} = \dfrac{n!}{(n-k)!k!}$
m	Index associated with a generic combination of k events	$m = 1,\ldots, N_k$
Ψ	Vessel vulnerability	See Tables 9.2–9.4
f	Overall expected frequency of the Natech scenario affecting the industrial facility	Evaluated according to specific models for the natural event of interest
$\delta(i, \boldsymbol{J}_m^k)$	Combination index	$\delta(i, \boldsymbol{J}_m^k) = 1$ if i-th event triggered by flooding belongs to the vector $\boldsymbol{J}_m^k$; $\delta(i, \boldsymbol{J}_m^k) = 0$ if not.
Evaluation of Combinations Probability and Frequency		
N_f	Number of different overall scenarios that may be generated by a single natural event	$N_f = \sum\limits_{k=1}^{n} \begin{pmatrix} n \\ k \end{pmatrix} = 2^n - 1$
$P_f^{(k,m)}$	Probability of occurrence of the m-th combination involving the simultaneous damage of k equipment	$P_f^{(k,m)} = \prod\limits_{i=1}^{n} \left[1 - \psi + \delta\left(i, \boldsymbol{J}_m^k\right)(2\psi - 1) \right]$
$f_f^{(k,m)}$	Frequency of occurrence of the m-th combination involving the simultaneous damage of k equipment	$f_f^{(k,m)} = f \cdot P_f^{(k,m)}$
Consequence Assessment Trough the Vulnerability Evaluation of Multiple Scenarios		
$V_{f,i}$	Vulnerability calculated for the (k,m) scenario triggered by Natech	
$V_f^{(k,m)}$	Vulnerability associated with the occurrence of the m-th combination involving the simultaneous damage of k equipment	$V_f^{(k,m)} = \min\left[\sum\limits_{i=1}^{m} V_{f,i}; 1 \right]$

Adapted from Antonioni et al. (2015).

9.1.4 Input Data and Calculation Procedure

The starting point of the procedure is the input of data on all the possible critical targets. The critical targets were defined in the present approach as all equipment items having a relevant inventory of hazardous substances (cf. Chapter 6). The GIS section of the ARIPAR-GIS software associates to a simplified layout (usually reporting only the equipment items and the main lines) the critical targets identified in the safety assessment of the plant. A single risk source is associated to each equipment item. In the Natech version, if the risk source is a possible target of the natural event, it may be associated to a vulnerability model, an equipment class, and to a secondary "Natech" event. Equipment involved only in Natech events, if present, may be represented by risk sources not associated to any primary scenario.

The default equipment classes in ARIPAR-GIS are atmospheric and pressurized tanks, elongated vessels, and auxiliary vessels, but further classes may be defined by the user. Specific equipment vulnerability models, yielding the equipment damage probability as a function of a severity vector used to quantify the severity of the natural event, are associated to each equipment class. Vulnerability models are usually defined in the software as probit equations, since this is the most common approach used in the literature. This issue will be further discussed in Section 9.1.5. Figure 9.2

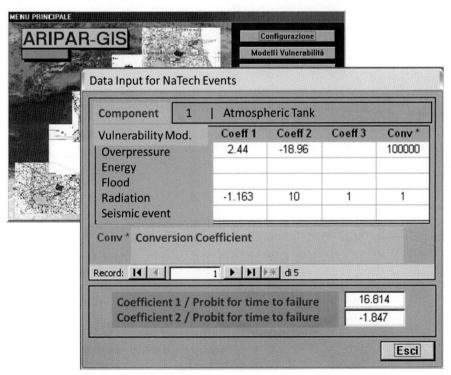

FIGURE 9.2 Management of Natech Equipment Vulnerability Models in the ARIPAR-GIS Software

shows an example of the equipment vulnerability model input interface provided by the software.

A secondary event is also associated to all the identified domino targets. A single secondary event is considered, in order to limit the computational effort required. Typically, the most severe credible scenario should be selected as the secondary event. An occurrence probability associated to this event is also required by the software. The occurrence probability represents the probability of the selected secondary event to take place given the equipment damage. The occurrence probability may be used to take into account that the equipment damage may not always be followed by a relevant secondary accident (e.g., the ignition of a release is not certain). In the absence of specific data, the occurrence probability should be conservatively assumed equal to 1.

The expected frequency of each domino scenario and the vulnerability map of each scenario are then calculated. The standard procedure of the ARIPAR-GIS software is used to estimate the contribution to the risk indices of all the identified domino scenarios.

9.1.5 Equipment Vulnerability Models

The detailed quantitative approach to the assessment of Natech scenarios presented in the previous section is based on the availability of models for equipment vulnerability. As mentioned in Chapter 7, several types of models may be applied to assess the failure probability of an equipment item due to the impact of a natural event. Detailed vulnerability models based on structural analysis may be developed, although these are time consuming and require unaffordable efforts in a QRA context. Observational fragility curves are also available in the literature (Salzano et al., 2003, 2009; Campedel et al., 2008; Antonioni et al., 2009).

In the framework of QRA, simple models are needed to allow the swift assessment of a high number of scenarios. Thus, fragility curves or simplified probabilistic models were selected for the ARIPAR-GIS software procedure for Natech risk assessment.

In the case of Natech accidents triggered by earthquakes, fragility curves expressed in the form of a probit function were selected:

$$Y = k_1 + k_2 \ln(\text{PGA}) \tag{9.1}$$

where PGA is the horizontal component of peak ground acceleration and the constants k_1 and k_2 are given in Table 9.2. With respect to floods, the equipment vulnerability models developed by Landucci et al. (2012, 2014) are compatible with the software. Tables 9.3 and 9.4 summarize these models which are based on the evaluation of the mechanical integrity of vessels under the action of the flood, which results in both a "static" external pressure component, due to the depth of the flooding, and in a "dynamic" external pressure component, due to the flood-water velocity and the associated kinetic energy. Considering this type of mechanical load,

Table 9.2 Values of the Probit Constants for Equipment Vulnerability Models Expressing Damage Probability Following an Earthquake

Type of Equipment	Damage State	Filling Level	$k_{1,i,j}$	$k_{2,i,j}$
Anchored atmospheric tanks	≥2	Near full	7.01	1.67
	≥2	≥50%	5.43	1.25
	3	Near full	4.66	1.54
	3	≥50%	3.36	1.25
Unanchored atmospheric tanks	≥2	Near full	7.71	1.43
	3	Near full	5.51	1.34
	3	≥50%	4.93	1.25
Horizontal pressurized storage tanks	≥1	Any	5.36	1.01
	≥2	Any	4.50	1.12
	3	Any	3.39	1.12
Pressurized reactors	≥1	Any	5.46	1.10
	≥2	Any	4.36	1.22
	3	Any	3.30	0.99
Pumps	≥2	—	5.31	0.77
	3	—	4.30	1.00

Different vulnerability models are provided as a function of equipment category, damage state, and filling level. Adapted from Campedel et al. (2008).

there is evidence that the vessel filling level is the most relevant parameter for the evaluation of the equipment integrity and associated fragility (i.e., its vulnerability). Thus, a critical filling level (CFL) was defined for each equipment item involved in a flood event of a specific intensity (e.g., having assigned flood-water velocity and depth) as the liquid level below which failure due to instability is possible. Further details are presented in Landucci et al. (2012, 2014).

9.1.6 Output

ARIPAR-GIS provides a number of different outputs. Local individual-risk maps allow a detailed mapping of the individual risk. If data on the population distribution is available, individual-risk maps for specific population categories (e.g., resident population, workers, etc.) can be calculated. This also includes vulnerability centers (i.e., sites where the aggregation of a high number of persons is expected, such as schools, hospitals, or railway stations.). ARIPAR-GIS outputs societal risk in the form of F–N curves or I–N diagrams. The latter is a measure of the exposure of society to the risk that plots the number of persons N in the impact area exposed to an individual risk within a specific range, I.

With the implementation of the Natech module, ARIPAR-GIS is currently the only software tool that provides a correct calculation of societal risk by being able

Table 9.3 Vulnerability Model and Input Parameters for Atmospheric Cylindrical Tanks Involved in Flood Events Based on the Critical Filling Level (CFL)

Item	Definition	Value/Equation
Vulnerability Model Equations		
CFL	Critical filling level	$CFL = \left(\dfrac{\rho_w k_w}{2} v_w^2 + \rho_w g h_w - P_{cr} \right) \Big/ \rho_t g H$
P_{cr}	Vessel critical pressure evaluated with the proposed simplified correlation	$P_{cr} = J_1 C + J_2$ in which $J_1 = -0.199$ $J_2 = 6950$
Ψ	Vessel vulnerability due to flooding	$\psi = \dfrac{CFL - \phi_{min}}{\phi_{max} - \phi_{min}}$
Input Parameters		
C	Vessel capacity	Small capacity C < 5,000 m³ Medium capacity 5,000–10,000 m³ Large capacity > 10,000 m³
v_w	Flood-water velocity[a]	0–3.5 m/s
h_w	Flood-water depth[a]	0–4 m
ρ_w	Flood-water density	1,100 kg/m³
ρ_t	Stored liquid density	650–1,300 kg/m³
k_w	Hydrodynamic coefficient	1.8
H	Vessel height	Small capacity 3.6–18 m Medium capacity 3.6–16.2 m Large capacity 3.6–7.2 m
g	Gravity acceleration	9.81 m/s²
ϕ_{min}	Minimum operative filling level	0.01
ϕ_{max}	Maximum operative filling level	0.75

[a]Parameters can be derived from the hydrogeological study of the analyzed area or provided by local competent authorities.
Adapted from Landucci et al. (2012).

to include the scenarios resulting from the simultaneous failure of more than one equipment item. This feature is not present in conventional software for QRA.

ARIPAR-GIS also allows the disaggregation of risk components, providing risk maps for specific scenarios or risk sources, thus allowing a sensitivity analysis and the identification of the most important risk sources and scenarios. Impact areas for the different scenarios considered can also be obtained. Further details on the output of the ARIPAR-GIS software are provided in Spadoni et al. (2000, 2003).

A quantitative risk analysis of earthquake and flood impacts at a hazardous installation using ARIPAR-GIS is presented in Chapter 11.

Table 9.4 Vulnerability Model and Input Parameters for Horizontal Cylindrical Tanks Involved in Flood Events Based on the Critical Filling Level (CFL)

Item	Definition	Value/Equation
Vulnerability Model Equations		
CFL_h	Critical filling level for horizontal vessels (pressurized or atmospheric)	$CFL_h = (\rho_{ref} \cdot A)/(\rho_l - \rho_v) \cdot (h_w - h_c)$ $+ (\rho_{ref} \cdot B - \rho_v)/(\rho_l - \rho_v)$
$v_{w,c}$	Flooding critical velocity	$v_{w,c} = E \cdot (h_w - h_c - h_{min})^F$
Ψ	Vessel vulnerability due to flooding	If $v_w \geq v_{w,c}$, $\Psi = 1$; If $v_w < v_{w,c}$, $\Psi = (CFL - \phi_{min})/(\phi_{max} - \phi_{min})$
Input Parameters		
C	Vessel capacity	Small capacity < 10 m³ Medium capacity 10–30 m³ Large capacity > 30 m³
W_t	Vessel tare weight[a]	900–2,200 kg (Small capacity) 3,000–7,200 kg (Medium capacity) 9,900–63,000 kg (Large capacity)
D	Vessel diameter	1.3–1.6 m (Small capacity) 1.6–2.4 m (Medium capacity) 2.3–3.8 m (Large capacity)
L	Vessel length	3–3.5 m (Small capacity) 4.5–11.1 m (Medium capacity) 8–24 m (Large capacity)
A	First CFL_h correlation coefficient	$A = K_1 \cdot D^a$
B	Second CFL_h correlation coefficient	$B = K_2 (W_t + K_3)^b$
E	$v_{w,c}$ correlation factor	$E = K_4 \cdot L^c$
F	$v_{w,c}$ correlation exponent	$F = K_5 \ln (L/D) + K_6$
K_1	Coefficient for A evaluation[a]	1.339
K_2	Coefficient for B evaluation[a]	−1.21
K_3	Coefficient for B evaluation[a]	−374.4
K_4	Coefficient for E evaluation[a]	5.497
K_5	Coefficient for F evaluation[a]	−0.06
K_6	Coefficient for F evaluation[a]	−0.375
a	Exponent for A evaluation[a]	−0.989
b	Exponent for B evaluation[a]	−0.107
c	Exponent for E evaluation	−0.692
v_w	Flood-water velocity[b]	0–3.5 m/s
h_w	Flood-water depth[b]	0–4 m
ρ_w	Flood-water density	1100 kg/m³
h_c	Height of concrete basement (flooding protection)	0.25 m
h_{min}	Minimum flooding height able to wet the vessel surface	$h_{min} = \lambda - D/2$

(Continued)

Table 9.4 Vulnerability Model and Input Parameters for Horizontal Cylindrical Tanks Involved in Flood Events Based on the Critical Filling Level (CFL) (cont.)

Item	Definition	Value/Equation
λ	Saddle height parameter which indicates the vessel axis height with respect to the ground anchorage point	0.98 m (Small capacity) 0.98–1.38 m (Medium capacity) 1.38–1.98 m (Large capacity)
ρ_l	Stored liquid density	500–1100 kg/m^3
ρ_v	Stored vapor density	1.25–20 kg/m^3
ρ_{ref}	Reference density used for the definition of CFL correlations	1000 kg/m^3
ϕ_{min}	Minimum operative filling level	0.01
ϕ_{max}	Maximum operative filling level	0.90

[a]*Value evaluated for 2 MPa design pressure.*
[b]*Parameters can be derived from the hydrogeological study of the analyzed area or provided by local competent authorities.*
Adapted from Landucci et al. (2014).

9.2 RISKCURVES

RISKCURVES is a computer program package that was developed by the Netherlands Organization for Applied Scientific Research (TNO) in the late 1980s to perform a QRA of hazardous activities due to conventional causal factors of accidents. The software has since then been upgraded continuously to introduce several innovative concepts, such as full integration of consequence modeling, showing societal risk on a map, calculation of risk contours and allowing external consequence data to be used for the risk calculation. Considering that a QRA is a complex task, special attention is paid to the user friendliness of the software and its capabilities to easily integrate the results in office- and GIS environments (TNO, 2015a).

RISKCURVES is a tool that aims to quantify the risk from the storage and transport of hazardous materials to the surrounding population and the built environment in urban areas and at chemical facilities. The risk sources include both fixed installations but also equipment used in the transport of hazardous materials (e.g., pipelines, road and rail tankers, ships). The tool allows the definition of a QRA with an unlimited number of fixed or transport equipment types with all their associated accident scenarios. Similar to RAPID-N, which was introduced in Section 8.1, RISKCURVES aims to support the user by offering different levels of user interaction during the assessment process, ranging from standard user on the one end in which user input is minimal, to expert user on the other end in which all input data is provided by the user him/herself (van het Veld et al., 2007). This means that depending on the complexity level, either RISKCURVES's internal assessment models will decide the analysis to various degrees or the user can fully customize the input data. This gives the user a maximum amount of flexibility in tailoring the assessment process, and it is a winning approach that has already shown its usefulness in the application of RAPID-N.

In order to support the consequence analysis, RISKCURVES includes the software package EFFECTS, also developed by TNO, which calculates the consequences of the accidental release of toxic and/or flammable chemicals. It includes models related to hazmat release, evaporation, and dispersion, as well as fire and explosion models (TNO, 2015b). Both RISKCURVES and EFFECTS are based on reference handbooks developed in the Netherlands, namely the *Yellow Book*, *Green Book*, *Purple Book*, and *Red Book* (VROM, 2005a,b,c,d), which are considered a standard reference for risk assessment by many risk-assessment practitioners.

The main output of RISKCURVES is individual and societal risk (*F–N* curves and societal-risk maps), as well as the consequence areas of the accident scenarios (Fig. 9.3). This includes the identification of the equipment and the scenarios that dominate the overall risk. By determining the area under the *F–N* curve, RISKCURVES also provides an estimate of the expected number of fatalities per year (van het Veld et al., 2007). This information can be used for risk management, decision-making, urban planning, and for any activity in support of compliance with criteria required by legislation.

RISKCURVES was developed for the analysis of conventional risks associated with hazardous activities. It therefore does not contain any specific models or

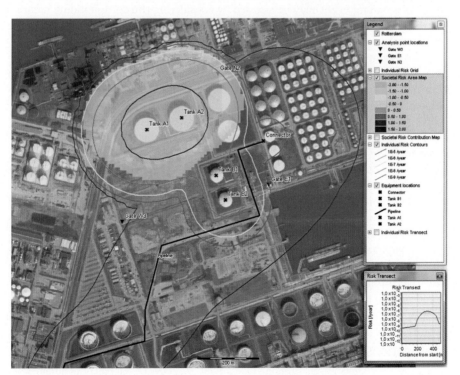

FIGURE 9.3 **Example of Iso-Risk Contours and Societal Risk Area Map for a QRA Using RISKCURVES**
(Courtesy: TNO)

software modules that explicitly take into account the interaction of natural events with industrial equipment or more generally Natech risks. However, this problem can be overcome by customizing the tool for Natech-type applications. This involves the use of models for equipment vulnerability analysis that consider natural-hazard intensities and recognizing a local probability of occurrence in terms of exceedance probability for the given intensity, as discussed, for example, in Chapters 5 and 7, or in Campedel et al. (2008) and Salzano et al. (2009).

For the Natech case study in Chapter 12, new source terms expressed as risk states (RSs) were introduced in RISKCURVES for each independent natural event with a given intensity measure IM, and for three equipment categories (atmospheric tank, pressurized vessel, and large pipes). These risk states are directly linked to the Natech fragility function P and to the natural hazard. The overall probability of exceeding a given RS was then defined as:

$$P[\text{RS} \geq \text{RS}_i] = \int_{\text{IM}} P[\text{RS} \geq \text{RS}_i \mid \text{IM}] \cdot h(\text{IM}) d\text{IM} \tag{9.2}$$

In other words, the RS probability of any equipment conditional to the occurrence of a natural event may be assessed by considering the corresponding hazard h of the natural event. The annual rate of RS exceedance is then calculated by using the annual rate of occurrence.

The fragility functions in Eq. (9.2) were defined in a similar way as in ARIPAR-GIS (Section 9.1.5). For earthquakes, the same analysis and functions as shown in Table 9.2 (Campedel et al., 2008) were adopted for different equipment. For the evaluation of the seismic fragility of pipes, the data and functions reported in Lanzano et al. (2014, 2015) were used. The only seismic intensity parameter considered was PGA.

For the analysis of tsunami-triggered Natech risk, the fragility functions as defined in Basco and Salzano (2016) were adopted. In this case, the main intensity parameter is the energy flux expressed in J/m^2 of the tsunami wave. This is equivalent to $\rho_w h_w v_w^2$ (where ρ, h, and v are the density, the height, and the velocity of the water wave) or, in other terms, the combination of kinetic and potential (i.e., buoyancy) energy of the wave. For tsunami debris, the Johnson number was considered (Corbett et al., 1996).

A complete overview of all features and further details of the software are available at www.tno.nl/riskcurves. Additional information is provided in Chapter 12 in which a customized version of RISKCURVES was applied to a QRA of an oil refinery located in the Mediterranean Sea under both earthquake and tsunami effects.

References

Antonioni, G., Bonvicini, S., Spadoni, G., Cozzani, V., 2009. Development of a framework for the risk assessment of Na-tech accidental events. Reliab. Eng. Syst. Saf. 94, 1442.

Antonioni, G., Landucci, G., Necci, A., Gheorghiu, D., Cozzani, V., 2015. Quantitative assessment of risk due to NaTech scenarios caused by floods. Reliab. Eng. Syst. Saf. 142, 334.

Antonioni, G., Spadoni, G., Cozzani, V., 2007. A methodology for the quantitative risk assessment of major accidents triggered by seismic events. J. Hazard. Mater. 147, 48.

Basco, A., Salzano, E., 2016. The vulnerability of industrial equipment to tsunami. J. Loss Prev. Process Ind., Accepted for publication.

Campedel, M., Cozzani, V., Garcia-Agreda, A., Salzano, E., 2008. Extending the quantitative assessment of industrial risks to earthquake effects. Risk Anal. 28, 1231.

Corbett, G., Reid, S., Johnson, W., 1996. Impact loading of plates and shells by free-flying projectiles: a review. Int. J. Impact Eng. 18, 141.

Cozzani, V., Antonioni, G., Landucci, G., Tugnoli, A., Bonvicini, S., Spadoni, G., 2014. Quantitative assessment of domino and NaTech scenarios in complex industrial areas. J. Loss Prev. Process Ind. 28, 10.

Cozzani, V., Antonioni, G., Spadoni, G., 2006. Quantitative assessment of domino scenarios by a GIS-based software tool. J. Loss Prev. Process Ind. 19 (5), 463.

Cozzani, V., Campedel, M., Renni, E., Krausmann, E., 2010. Industrial accidents triggered by flood events: analysis of past accidents. J. Hazard. Mater. 175, 501.

Cozzani, V., Gubinelli, G., Antonioni, G., Spadoni, G., Zanelli, S., 2005. The assessment of risk caused by domino effect in quantitative area risk analysis. J. Hazard. Mater. 127 (1–3), 14.

Egidi, D., Foraboschi, F.P., Spadoni, G., Amendola, A., 1995. The ARIPAR project: analysis of the major accident risks connected with industrial and transportation activities in the Ravenna area. Reliab. Eng. Syst. Saf. 49 (1), 75.

Landucci, G., Antonioni, G., Tugnoli, A., Cozzani, V., 2012. Release of hazardous substances in flood events: damage model for atmospheric storage tanks. Reliab. Eng. Syst. Saf. 106, 200.

Landucci, G., Necci, A., Antonioni, G., Tugnoli, A., Cozzani, V., 2014. Release of hazardous substances in flood events: damage model for horizontal cylindrical vessels. Reliab. Eng. Syst. Saf. 132, 125.

Lanzano, G., Salzano, E., Santucci de Magistris, F., Fabbrocino, G., 2014. Seismic vulnerability of gas and liquid buried pipelines. J. Loss Prev. Process Ind. 28, 72.

Lanzano, G., Santucci de Magistris, F., Fabbrocino, G., Salzano, E., 2015. Seismic damage to pipelines in the framework of Na-Tech risk assessment. J. Loss Prev. Process Ind. 33, 159.

Reniers, G., Cozzani, V., 2013. Domino Effects in the Process Industries: Modeling, Prevention and Managing. Elsevier, Amsterdam.

Renni, E., Krausmann, E., Cozzani, V., 2010. Industrial accidents triggered by lightning. J. Hazard. Mater. 184, 42.

Salzano, E., Garcia-Agreda, A., Di Carluccio, A., Fabbrocino, G., 2009. Risk assessment and early warning systems for industrial facilities in seismic zones. Reliab. Eng. Syst. Saf. 94, 1577.

Salzano, E., Iervolino, I., Fabbrocino, G., 2003. Seismic risk of atmospheric storage tanks in the framework of quantitative risk analysis. J. Loss Prev. Process Ind. 16, 403.

Spadoni, G., Contini, S., Uguccioni, G., 2003. The new version of ARIPAR and the benefits given in assessing and managing major risks in industrialized areas. Process Saf. Environ. Protect. 81 (1), 19.

Spadoni, G., Egidi, D., Contini, S., 2000. Through ARIPAR-GIS the quantified area risk analysis supports land-use planning activities. J. Hazard. Mater. 71 (1–3), 423.

TNO, 2015a. RISKCURVES, Risk assessment software, Version 9. The Netherlands Organization for Applied Scientific Research (TNO). www.tno.nl/riskcurves

TNO, 2015b. EFFECTS, Software for safety- and hazard analysis. The Netherlands Organization for Applied Scientific Research (TNO). www.tno.nl/effects/

van het Veld, F., Boot, H., Kootstra, F., 2007. RISKCURVES: a comprehensive computer program for performing a Quantitative Risk Assessment (QRA). In: Aven, T., Vinnem, J.E. (Eds.), Risk, Reliability and Societal Safety. Taylor & Francis Group, London, p. 1547.

VROM, 2005a. Methods for the determination of possible damage to people and objects resulting from releases of hazardous materials. Green Book, PGS 1. Ministry of Housing, Spatial Planning and the Environment, The Hague.

VROM, 2005b. Guidelines for quantitative risk assessment. Purple Book, PGS 3. Ministry of Housing, Spatial Planning and the Environment, The Hague.

VROM, 2005c. Methods for determining and processing probabilities. Red Book, PGS 4. Ministry of Housing, Spatial Planning and the Environment, The Hague.

VROM, 2005d. Methods for the calculation of physical effects. Yellow Book, PGS 2. Ministry of Housing, Spatial Planning and the Environment, The Hague.

Case-Study Application I: RAPID-N

S. Girgin, E. Krausmann
European Commission, Joint Research Centre, Ispra, Italy

In this chapter, the rapid Natech risk analysis and mapping framework RAPID-N introduced in Chapter 8 is used to carry out a simplified Natech risk analysis for an industrial facility in Izmit Bay in Turkey that was subjected to a predicted Istanbul earthquake scenario. The results demonstrate RAPID-N's capability to assess the earthquake impact on an industrial plant, including the simultaneous analysis of the Natech risk at several plant units.

10.1 EARTHQUAKE SCENARIO

The Marmara region is one of the most tectonically active regions in Eurasia. Over the last century, seismic activity with nine earthquakes with $M_w \geq 7$ was registered. The Kocaeli and Düzce earthquakes in 1999 were two extremely destructive events that occurred in the eastern part of the region along the North Anatolian Fault (NAF). The NAF is a strike-slip fault system that crosses the north of Turkey for over 1200 km and accommodates about 25 mm right lateral slip per year between the Anatolian and the Eurasian plate (McClusky et al., 2000; Straub et al., 1997).

The large earthquakes generated by the NAF are in a sequence that appears to propagate westward (Stein et al., 1997; Barka, 1992; Ambraseys, 1970). The 1999 Kocaeli earthquake occurred in the southern part of the eastern border of Istanbul province. The westward motion of the earthquakes suggests that Istanbul is at high risk of being hit by strong future seismic activity. Studies estimate that the occurrence probability of $M_w \geq 7$ earthquakes in the Marmara region which could impact the Istanbul Metropolitan area is $41 \pm 14\%$ for the time period of 2004–34 (Parsons, 2004). The Yalova fault segment in the south of Istanbul and the Northern Boundary fault in the southeast have the potential to rupture and are therefore of the biggest concern in this context (Parsons et al., 2000; Hubert-Ferrari et al., 2000).

The high level of seismic risk warrants an in-depth assessment of the regional earthquake hazard to understand potential impacts to the urban area and the hazardous industry it includes. The Istanbul disaster prevention and mitigation plan completed in 2002 considers four different scenario earthquakes for the assessment of potential seismic damage (JICA, 2002). These four scenarios differ in the assumed location and length of NAF rupture. For this RAPID-N case study we have selected

Natech Risk Assessment and Management. http://dx.doi.org/10.1016/B978-0-12-803807-9.00010-3

the Model A scenario earthquake, which is characterized by an M_w = 7.5 seismic event that results in the rupture of a more than 120 km strike-slip fault from west of the 1999 Kocaeli earthquake fault. This scenario is considered the most probable scenario in the disaster prevention and mitigation plan for Istanbul.

RAPID-N can in principle calculate ground-motion parameters at the location of hazardous installations by using scenario-earthquake parameters and available ground-motion prediction equations. However, for consistency with previous studies it was decided to use precalculated on-site data provided by AFAD, the Turkish Prime Ministry Disaster and Emergency Management Presidency (AFAD, personal communication). For this purpose, a 0.02° × 0.02° data grid covering the Marmara region that included peak ground acceleration (PGA), peak ground velocity (PGV), and earthquake intensity (MMI) values was converted into a ShakeMap supported by RAPID-N. Precalculated ground-motion parameters were based on the attenuation equation by Boore et al. (1997) and Campbell (1997) (200% of the estimate) for PGA and PGV, respectively. They were corrected for subsurface amplification according to Wald et al. (1999), considering on-site soil classes from the National Earthquake Hazards Reduction Program (NEHRP).

The scenario earthquake with the assumed fault rupture and the corresponding regional PGA estimates is shown in Fig. 10.1. As indicated in the Fig. 10.1, if

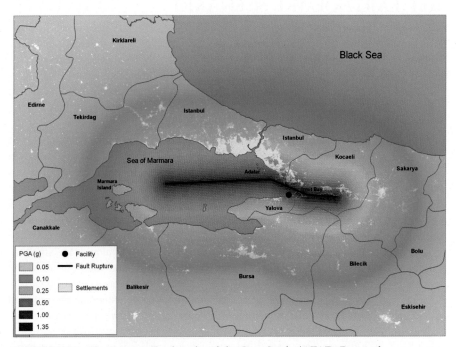

FIGURE 10.1 The Scenario Earthquake of the Case Study (AFAD, Personal Communication)

The *black dot* denotes the location of the case study installation.

the scenario earthquake occurred, the area around Izmit Bay would be affected the most, with PGA values of up to 1.35 g. On the northern (Körfez) and southern (Tasköprü) shores of Izmit Bay, PGA values of 0.75 g are predicted. Lower PGAs of about 0.4 g are expected on the seashore of the European side of Istanbul and Adalar, whereas forecasts for the Asian side of Istanbul estimate a PGA of less than 0.3 g. The maximum predicted PGV and MMI are 2.7 m/s and 10.8, respectively. MMI values greater than 10 correspond to significant destruction to man-made structures.

10.2 CHEMICAL FACILITY DESCRIPTION

Several hazardous industrial installations in Izmit Bay were damaged during the 1999 Kocaeli earthquake (Girgin, 2011; Durukal and Erdik, 2008; Suzuki, 2002; Steinberg and Cruz, 2004; Rahnama and Morrow, 2000). Considering that the Bay area is predicted to exhibit the highest ground motion values for the Istanbul scenario earthquake, we selected an industrial plant located on the southern shore of Izmit Bay (Fig. 10.1). This installation, which is operational since 1971, produces acrylic textile and technical fibers with a production capacity of 315,000 tons/year in 2015. The installation includes a harbor, a feedstock storage tank farm, and facilities for polimerization, DOP production, solvent recovery, fiber pullout, cutting packaging, product storage, and waste-water treatment. The site also hosts a carbon-fiber production facility with a capacity of 3500 tons/year and a large coal/natural-gas hybrid power plant with a total capacity of 142.5 MW/year including coal storage silos. The site is surrounded by Izmit Bay in the north and crop lands in the west and east with a limited number of residential settlements. On the south there is a major road along which other industrial facilities are located (Fig. 10.2).

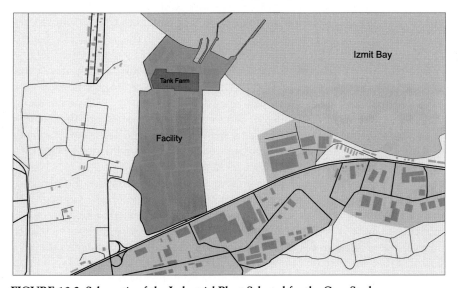

FIGURE 10.2 Schematic of the Industrial Plant Selected for the Case Study

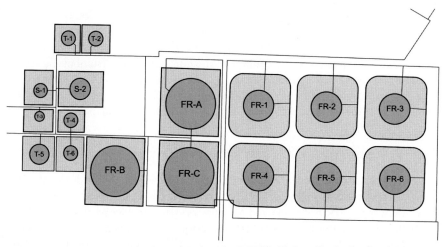

FIGURE 10.3 Storage Tanks Considered in the RAPID-N Case Study Application

The installation suffered severe damage to three storage tanks during the Kocaeli earthquake in 1999, which was of similar magnitude as the expected Istanbul earthquake. About 6500 tons of acrylonitrile were released into the air, sea, and groundwater, and all animals and vegetation were lethally affected inside the facility within a 200-m radius around the tanks. Acute toxicity symptoms were observed in emergency response teams and residents in the vicinity (Girgin, 2011). There is, therefore, an opportunity to compare the outcome of our Natech risk analysis with historical data.

As a first step in the Natech risk analysis with RAPID-N, the selected plant was identified in a high-resolution satellite image, its boundaries were delineated, and the storage tanks located and mapped. This was done with the user-friendly mapping tool built into the RAPID-N framework. Overall, 17 storage tanks were identified on site (Fig. 10.3). Tank shapes (e.g., cylindrical vertical, spherical), planar dimensions, foundation types (e.g., on-ground, elevated), as well as the presence and size of containment dikes were also identified using the satellite image. Additional satellite images that were acquired at different dates were utilized to determine the roof types (e.g., fixed, floating roof). Once the tank shape and roof type had been determined, the substance storage conditions were deduced. A summary of the storage tank characteristics defined in this way is shown in Table 10.1.

Although satellite images provide adequate information about the planar dimensions of storage tanks, vertical dimensions are difficult to obtain. Consequently, the tank heights and the depth of the surrounding containment dikes were not directly measurable. Therefore, it was not possible to calculate the exact tank and dike volumes. Although RAPID-N includes tank dimensions and volume estimation functions that are based on common design codes and typical storage tanks, additional data were collected from a public document published by the facility to improve the analysis. By matching satellite imagery-based information to the information

Table 10.1 Storage Tank Characteristics Determined From Satellite Imagery

Unit ID	Tank Shape	Roof Type	Foundation Type	Storage Condition	Diameter (m)	Dike Area (m × m)
FR-A	Cylindrical	Internal floating	On-ground	Atmospheric	42.5	50 × 55
FR-B	Cylindrical	External floating	On-ground	Atmospheric	42.5	50 × 55
FR-C	Cylindrical	External floating	On-ground	Atmospheric	42.5	50 × 55
FR-1	Cylindrical	Internal floating	On-ground	Atmospheric	25.0	50 × 50
FR-2	Cylindrical	Internal floating	On-ground	Atmospheric	25.0	50 × 50
FR-3	Cylindrical	Internal floating	On-ground	Atmospheric	25.0	50 × 50
FR-4	Cylindrical	Internal floating	On-ground	Atmospheric	25.0	50 × 50
FR-5	Cylindrical	Internal floating	On-ground	Atmospheric	25.0	50 × 50
FR-6	Cylindrical	Internal floating	On-ground	Atmospheric	25.0	50 × 50
S-1	Spherical	—	Elevated	Pressurized	12.5	32 × 30
S-2	Spherical	—	Elevated	Pressurized	18.0	42 × 30
T-1	Cylindrical	Fixed	On-ground	Atmospheric	12.2	24 × 24
T-2	Cylindrical	Fixed	On-ground	Atmospheric	12.2	24 × 24
T-3	Cylindrical	Fixed	On-ground	Atmospheric	7.6	30 × 22
T-4	Cylindrical	Fixed	On-ground	Atmospheric	12.5	25 × 22
T-5	Cylindrical	Fixed	On-ground	Atmospheric	17.2	30 × 30
T-6	Cylindrical	Fixed	On-ground	Atmospheric	12.5	25 × 30

available in the document, the substances stored in each tank and their corresponding dike volumes were also determined.

Of the analyzed tanks, five are atmospheric internal floating roof tanks which contain acrylonitrile. Acrylonitrile is a raw material used for the production of acrylic fibers that is classified as highly flammable, toxic, hazardous to the aquatic environment, and it may cause cancer (EU, 2008; IARC, 1999). Two spherical tanks contain ammonia, whereas the six fixed-roof tanks contain acetic acid, vinyl acetate, and methanol, all of which are flammable and toxic chemicals.

No information was available on the type of substances stored in the remaining two external floating roof tanks. Consequently, they were excluded from the Natech risk analysis as no consequence analysis is possible without substance data. In fact, in the satellite image these tanks appear to have a very low fill level and hence they were considered empty for the purpose of the case study. All tanks were assumed to be anchored since the facility is located in a highly seismic area and had Natech experience in the past. Similarly, support columns of the spherical pressurized tanks were assumed to be braced diagonally for better seismic performance.

In addition to the overall storage tank capacity, the amount of hazardous materials actually present is also a determining factor for the risk analysis. In order to simulate actual operating conditions and also to demonstrate the effect of fill level, which is an important factor in determining the degree of structural damage of storage tanks, we assumed different fill levels corresponding to near-full and half-full

Table 10.2 Type and Amount of Substances in the Tanks Considered in the Case Study

Unit ID	Stored Substance	Height (m)	H/D Ratio	Capacity (m³)	Dike Volume (m³)	Fill Level (%)	Stored Quantity (tons)
FR-A	Acrylonitrile	10.5	0.25	16,000	16,000	50	6,366
FR-1	Acrylonitrile	10.5	0.42	5,042	5,050	60	2,407
FR-2	Vinyl acetate	10.5	0.42	5,044	5,050	60	2,772
FR-3	Vinyl acetate	10.5	0.42	5,044	5,050	40	1,848
FR-4	Acrylonitrile	10.5	0.42	5,042	5,050	60	2,407
FR-5	Acrylonitrile	10.5	0.42	5,042	5,050	50	2,006
FR-6	Acrylonitrile	10.5	0.42	5,042	5,050	40	1,605
S-1	Ammonia	12.5	1.00	972	1,270	80	532
S-2	Ammonia	18.0	1.00	3,002	1,800	80	1,642
T-1	Methanol	9.0	0.74	1,059	577	80	671
T-2	Methanol	9.0	0.74	1,060	577	40	336
T-3	Acetic acid	11.0	1.45	504	750	80	423
T-4	Acetic acid	9.0	0.72	1,087	1,360	40	456
T-5	Methanol	9.5	0.55	2,186	882	80	1,385
T-6	Acetic acid	9.0	0.72	1,080	1,360	60	680

conditions for various tanks. Table 10.2 gives a summary of the tanks included in the RAPID-N risk-analysis case study, the substances they contain, reported storage capacities, dike volumes, and the assumed fill levels.

10.3 NATECH RISK ANALYSIS

For this Natech risk analysis with RAPID-N, the information on the severity of the Istanbul earthquake scenario at the location of the selected hazardous installation was used to assess the predicted damage to the plant's storage tanks and its likelihood. Based on the damage analysis, three different types of analysis were carried out to understand the earthquake impact on (1) a storage tank containing a flammable substance, (2) a storage tank containing a toxic substance, and (3) the multiple plant units listed in Table 10.2.

10.3.1 Damage Analysis

The Istanbul earthquake scenario provided by AFAD served as natural-hazard scenario for all three case-study applications. The associated ShakeMap was used to determine the on-site earthquake severity needed for the risk analysis. For this purpose, RAPID-N calculated the distance of each storage tank from the epicenter and interpolated the seismic hazard parameters from the ShakeMap. The storage tanks

Table 10.3 Damage Classifications Utilized for the Case Study

State	O'Rourke and So (2000)	Moschonas et al. (2014)
DS1	No damage to tank or I/O pipes.	No damage.
DS2	Damage to roof, minor loss of contents, minor damage to piping, but no elephant-foot buckling.	Minor yields that correspond to minor permanent deformations at critical sections of a small percentage of columns and/or braces.
DS3	Elephant-foot buckling with minor loss of content.	Moderate yields corresponding to moderate permanent deformations at critical sections of a moderate percentage of columns and/or braces without any global buckling failure of columns.
DS4	Elephant-foot buckling with major loss of content, severe damage.	Major yields causing major permanent deformations at critical sections of a large percentage of columns and/or braces with global buckling failure of columns where maximum compression occurs.
DS5	Total failure, tank collapse.	Buckling failure with subsequent collapse of the pressure vessel.

lie at a distance of 6.1–6.3 km from the predicted location of the epicenter. Peak ground acceleration and velocity are of the order of 0.8 g and 1.7 m/s, respectively. The ShakeMap also provides data on instrumental earthquake intensity, which is expected to exceed 10, corresponding to *very destructive* on the European Macroseismic Intensity Scale (Grünthal, 1998). This suggests that if this earthquake were to occur, many buildings designed according to current standards would suffer damage or even collapse, were they located at the site of the chemical installation.

In its current version, RAPID-N includes the most frequently used damage classifications and fragility curves for storage tanks available in the scientific literature. For this case study, RAPID-N automatically selected for each plant unit a fragility curve and a corresponding damage classification that represented the best fit with the available data. Where plant-unit characteristics required for the calculations were missing, these data were estimated using RAPID-N's built-in property estimation framework. For the 15 storage tanks included in the damage analysis, RAPID-N utilized the fragility curves defined by O'Rourke and So (2000) for atmospheric cylindrical tanks, and those of Moschonas et al. (2014) for pressurized spherical tanks. The different damage states used and their definitions are summarized in Table 10.3. The damage states of O'Rourke and So are similar to HAZUS damage states commonly used for seismic damage assessment (FEMA, 2010).

O'Rourke and So (2000) provide four different fragility curve sets for atmospheric storage tanks applicable for *fill levels* <50% and ≥50%, and *height/diameter* (H/D) ratios <0.7 and ≥0.7. For pressurized tanks, Moschonas et al. (2014) provide two fragility curve sets for column support *with* and *without* diagonal braces. Using this information, three different fragility curves were selected by RAPID-N to assess the

Table 10.4 Summary of Damage Parameters for the Earthquake Damage Analysis

Unit ID	Distance[a] (km)	PGA (g)	PGV (m/s)	Fragility Curve	Damage Probability (%)				
					DS1	DS2	DS3	DS4	DS5
FR-A	6.24	0.78	1.67	O'Rourke and So, H/D < 0.7	37.7	50.9	9.0	2.2	0.2
FR-1	6.25	0.78	1.68	O'Rourke and So, H/D < 0.7	37.6	50.9	9.1	2.3	0.2
FR-2	6.26	0.79	1.68	O'Rourke and So, H/D < 0.7	37.4	50.9	9.1	2.3	0.2
FR-3	6.26	0.79	1.68	O'Rourke and So, H/D < 0.7	37.3	51.0	9.2	2.3	0.2
FR-4	6.31	0.78	1.67	O'Rourke and So, H/D < 0.7	38.0	50.8	8.8	2.2	0.2
FR-5	6.31	0.78	1.67	O'Rourke and So, H/D < 0.7	37.9	50.8	8.9	2.2	0.2
FR-6	6.32	0.78	1.67	O'Rourke and So, H/D < 0.7	37.7	50.9	9.0	2.2	0.2
S-1	6.22	0.78	1.67	Moschonas et al., Braced	8.7	84.0	6.9	0.3	0.01
S-2	6.22	0.78	1.67	Moschonas et al., Braced	8.7	84.0	6.9	0.3	0.01
T-1	6.18	0.78	1.68	O'Rourke and So, H/D ≥ 0.7	11.8	22.5	38.2	25.5	1.9
T-2	6.18	0.79	1.68	O'Rourke and So, H/D ≥ 0.7	11.8	22.5	38.1	25.6	2.0
T-3	6.25	0.78	1.67	O'Rourke and So, H/D ≥ 0.7	12.1	23.1	38.5	24.6	1.7
T-4	6.25	0.78	1.67	O'Rourke and So, H/D ≥ 0.7	12.1	23.0	38.4	24.7	1.7
T-5	6.28	0.78	1.66	O'Rourke and So, H/D < 0.7	38.3	50.7	8.6	2.1	0.2
T-6	6.28	0.78	1.66	O'Rourke and So, H/D ≥ 0.7	12.2	23.2	38.5	24.4	1.7

[a]From the epicenter.

probability of the different damage categories. The curves and the corresponding damage probabilities as a function of damage state are summarized in Table 10.4. The damage parameters are provided for discrete rather than cumulative damage categories and show the probability of a certain level of damage as shown in Fig. 10.4.

The damage data in Table 10.4 were subsequently used in the Natech risk analysis. As outlined in Chapter 8, the expected damage states and their probabilities were linked to consequence scenarios via *risk state* definitions, and the consequences were then estimated by using scenario-specific, dynamically generated consequence models formed by RAPID-N. For some damage states, the defined physical damage does not

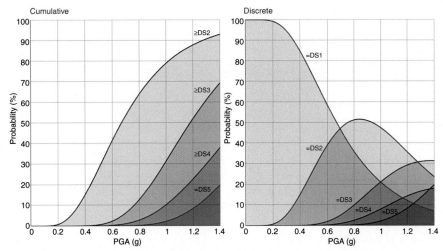

FIGURE 10.4 Example Cumulative and Discrete Fragility Curves [O'Rourke and So (2000), H/D < 0.7]

imply loss of containment. For some others, no explicit information is provided by the damage definition about substance amount possibly released although physical damage leading to a release is indicated. Since historical data about release rates due to seismic damage are scarce, possible release scenarios were defined based on expert judgment as percentage of tank volume released for each damage state. Similarly, the conditional probability of release given a specific degree of physical damage was assigned to each damage state since loss of containment is not always expected, especially if physical damage is limited. Risk states used for the case study are summarized in Table 10.5.

Since the consequence-analysis methods and equations currently available in RAPID-N are based on US EPA's Risk Management Program (RMP) guidance for offsite consequence analysis methodology (US EPA, 1999), a number of simplifying assumptions were made for the risk analysis. For example, for all scenarios the atmospheric stability was assumed to be neutral (corresponding to Pasquill stability class D) with a wind speed of 3 m/s, which is the *alternative scenario* according to RMP. The direction of the wind was not taken into account as it is not included in the RMP methodology. The ambient temperature was assumed to be 25°C and the

Table 10.5 Summary of Risk States Used for the Case Study

State	O'Rourke and So (2000)	Moschonas et al. (2014)
DS1	No release	No release
DS2	2% release, 30% release probability	No release
DS3	5% release, 50% release probability	2% release, 60 min, 50% release probability
DS4	50% release, 80% release probability	20% release, 60 min, 80% release probability
DS5	100% release, 100% release probability	100% release, 10 min, 100% release probability

Table 10.6 *Summary of Chemical Properties of Hazardous Substances Used for the Case Study*

Property	Methanol	Vinyl Acetate	Acetic Acid	Ammonia	Acrylonitrile
CAS No.	67-56-1	108-05-4	64-19-7	7664-41-7	107-13-1
EC No.	200-659-6	203-545-4	200-580-7	231-635-3	203-466-5
Chemical formula	CH_3OH	$C_4H_6O_2$	$C_2H_4O_2$	NH_3	C_3H_3N
Molecular weight (g/mol)	32.04	86.09	60.05	17.03	53.06
Density (g/cm³)	0.792	0.934	1.049	0.682	0.806
Boiling point (°C)	64.7	72.5	118.5	−33.3	77.3
Vapor pressure (mmHg)	126.9	115.0	15.9	7524.0	106.3
Vapor density	1.11	2.97	2.07	0.59	1.83
Flash point (°C)	11	−8	39	—	−1
Lower explosive limit (%)	6	2.6	4	15	3
Upper explosive limit (%)	36.5	13.4	16	28	17
Heat capacity (J/mol·K)	81.1	165.0	123.1	35.1	110.9
Heat of vaporization (kJ/mol)	39.2	34.4	52.3	23.3	31.8
Heat of combustion (kJ/mol)	723	1931	873	316	1718
ERPG-2 concentration (ppm)	1000	75	35	150	35

humidity 50%. For surface roughness, rural topography (flat terrain) was considered as the facility is mainly surrounded by sea and crop fields as shown in Fig. 10.2. Hazmat releases due to the seismic forces were assumed to occur at ground level. In addition, for unbounded liquid spills (i.e., not contained within a dike or in case of dike overflow) a minimum pool depth of 10 cm created by a substance release was assumed. While the RMP methodology adopts a pool depth of 1 cm, it was found to yield unrealistic pool sizes, in particular if the amount of released substance is high. For relative probabilities of ignition and explosion for flammable substances, estimates by Cox et al. (1990) were used. The chemical properties of the hazardous materials used for the consequence analysis are summarized in Table 10.6.

10.3.2 Single Unit Containing a Flammable Substance

For analyzing the Natech risk originating from the storage of a flammable substance, atmospheric fixed-roof storage tank T-1 containing methanol with an assumed 80% fill level was studied. According to the results of the damage analysis for which the O'Rourke and So (2000) H/D ≥ 0.7 fragility curve was used, the most likely consequence scenario is pool fire for all damage states except DS1 which is characterized by no damage. No part of the cloud formed by evaporation of the released methanol was found to be above the lower explosive limit, therefore explosion was excluded as an outcome of the analysis. The pool fire end-point distances were calculated using the TNO single point model (TNO, 1996). An end-point radiation intensity of 5 kW/m² corresponding to second degree burns if exposed for 40 s was assumed, which is the standard according to RMP. At this intensity, emergency actions lasting up to several minutes

Table 10.7 RAPID-N Output for Earthquake Impact on Tank T-1 Containing Methanol

State	Consequence Scenario	End-Point Distance (m)	Natech Probability (%)
DS1	No release	—	—
DS2	16.9 m³ release; 459 m² pool (within dike)	33.5	0.07
DS3	42.4 m³ release; 459 m² pool (within dike)	33.5	0.19
DS4	423.6 m³ release; 459 m² pool (within dike)	33.5	0.20
DS5	847.2 m³ release; 3161 m² pool (dike overflow)	88.0	0.02

may be conducted without shielding but with protective clothing (API, 1990). RAPID-N calculates a minimum end-point distance of 33.5 m with an occurrence probability of 6.7×10^{-4}, and a maximum end-point distance corresponding to the worst-case damage state of 88 m and a probability of 1.9×10^{-4}. Table 10.7 summarizes the output of the RAPID-N consequence analysis. The associated impact area map with end-point distances is shown in Fig. 10.5. The darker areas highlight impact zones with a higher probability of damage and loss.

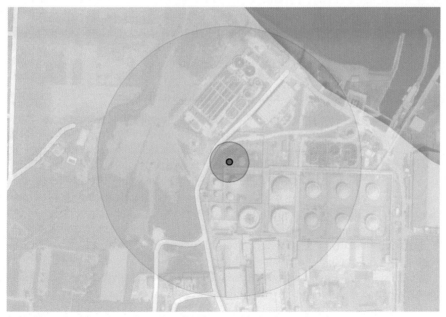

FIGURE 10.5 Natech Impact Zone for Heat Radiation From Tank T-1 Containing Methanol (Base Image ©2016 DigitalGlobe)

For damage states DS2 and DS3, some methanol is predicted to be released which forms a pool that ignites but stays confined in the tank's dike area. In the case of DS4, the amount of released substance increases significantly. However, because the dike's storage capacity is big enough to hold the methanol spill, the end-point distance does not increase and the thermal effects of the pool fire stay mitigated. For the worst-case damage scenario DS5, the whole tank volume is released due to the earthquake damage. The dike cannot hold the spilled methanol, causing a dike overflow and the spreading of the methanol beyond the dike perimeter. Assuming that the spill creates an unbounded pool with a minimum depth of 10 cm outside the dike area, the resulting pool area increases more than twofold. Consequently, the end-point distance for heat radiation also shows a substantial increase.

Fig. 10.5 indicates that depending on the level of damage to T-1, some other tanks on-site are predicted to fall within the heat radiation zones. Since the thermal-radiation intensity criteria are defined for humans, escalation effects (e.g., domino accidents) are not expected at this level, although some physical damage to other units is probable. Heat intensities that can significantly affect storage tanks are in the range of 9.5–38 kW/m², and intensity limits suggested for the quantitative risk analysis of domino effects are 15 and 45 kW/m² for atmospheric and pressurized storage tanks, respectively (Cozzani et al., 2006). Additional analyses with RAPID-N for these end-point intensities showed that among the tanks adjacent to T-1, the pressurized storage tanks containing ammonia (S-1 and S-2) are located outside the 45 kW/m² end-point distance estimated as 11 m for DS2–DS4 and 29 m for DS5. However, the atmospheric storage tank containing methanol (T-2) is inside the 15 kW/m² end-point distance calculated as 19 m for DS2–DS4 and 51 m for DS5. It is also adjacent to the higher 45 kW/m² end-point distance. RAPID-N therefore gives an indication of the other plant units potentially at risk from heat impingement if T-1 undergoes damage and the released methanol ignites. While not providing a detailed quantitative estimate of this risk, RAPID-N can nonetheless highlight areas of concern due to potential domino effects.

10.3.3 Single Unit Containing a Toxic Substance

For analyzing the Natech risk originating from a toxic substance, we used stage tank FR-1 which is an atmospheric internal floating roof tank containing acrylonitrile that is anchored and 60% full. The O'Rourke and So (2000) H/D < 0.7 fragility curve was used in the damage analysis. The analysis indicates that the most likely consequence scenario is the dispersion of the toxic substance in the atmosphere for all damage states with release. The reference toxic concentration used for calculating the end-point distances is the ERPG-2 concentration of 0.076 mg/L (35 ppm). It is the maximum airborne concentration below which nearly all individuals could be exposed for up to 1 h without experiencing or developing irreversible or other serious health effects or symptoms that could impair an individual's ability to take protective action (AIHA, 1988).

A dense-plume model was used to simulate the atmospheric dispersion of acrylonitrile, which is heavier than air. Reference Table 19 (dense gas, 60-min release, rural conditions, atmospheric stability D, wind speed 3 m/s) of the RMP guidance was utilized to determine the end-point distances. In accordance with the definition

Table 10.8 RAPID-N Output for Earthquake Impact on Tank FR-1 Containing Acrylonitrile

State	Consequence Scenario	End-Point Distance (km)	Natech Probability (%)
DS1	No release	—	—
DS2	60.5 m³ release; 2009 m² pool (within dike)	2.4	15.3
DS3	151.3 m³ release; 2009 m² pool (within dike)	2.4	4.5
DS4	1512.6 m³ release; 2009 m² pool (within dike)	2.4	1.8
DS5	3025.2 m³ release; 2009 m² pool (within dike)	2.4	0.2

of damage state DS1, no release occurs although there is some seismic loading. For damage states DS2–DS5, RAPID-N predicts the release of acrylonitrile and evaporating pool formation, which results in an end-point distance of 2.4 km with occurrence probabilities ranging between 1.5×10^{-1} and 2.2×10^{-3}. The released amount of substance shows a pronounced increase for DS4 and DS5, but the end-point distance is not affected. Similar to the analysis for tank T-1, the dike's holding capacity is sufficient to keep the substance confined within the dike, this time even for the worst-case scenario, and hence evaporation is limited. This demonstrates the importance of measures to contain the spill. The results of the analysis for tank FR-1 are summarized in Table 10.8 while the impact areas are shown in Fig. 10.6.

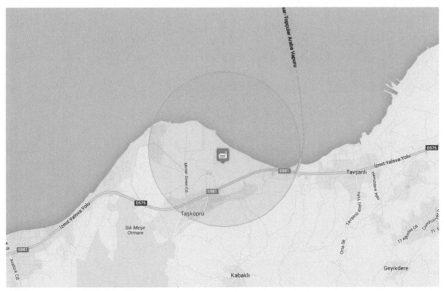

FIGURE 10.6 Natech Impact Zone for Atmospheric Dispersion of Acrylonitrile From Tank FR-1 (Map Data ©2016 Google)

10.3.4 Multiple Units

It is a characteristic of Natech accidents that multiple hazardous-materials releases from different plant units often occur at the same time. RAPID-N takes this characteristic into account and provides a framework that also allows the analysis of this aspect of Natech risk. Using the case-study installation, RAPID-N analyzed and mapped the potential for simultaneous damage to multiple plant units. For this purpose, all tanks for which data on the stored substances were available were included in the analysis (Table 10.2). This concerned four storage tanks with acrylonitrile, two with vinyl acetate, two with ammonia, three with methanol, and three containing acetic acid. The output of the analysis is summarized in Table 10.9 while the impact zones are shown in Fig. 10.7.

Table 10.9 Summary of the Natech Risk Assessment Results for All Tanks

Unit ID	Event	DS2	DS3	DS4	DS5
FR-A	Toxic dispersion	160 m³, 1.6 km, 15.3%	400 m³, 1.6 km, 4.5%	4000 m³, 1.6 km, 1.8%	8000 m³, 1.6 km, 0.2%
FR-1	Toxic dispersion	60.5 m³, 2.4 km, 15.3%	151.3 m³, 2.4 km, 4.5%	1512.6 m³, 2.4 km, 1.8%	3025.2 m³, 2.4 km, 0.2%
FR-2	Toxic dispersion	60.5 m³, 1.6 km, 15.3%	151.3 m³, 1.6 km, 4.6%	1512.6 m³, 1.6 km, 1.8%	3025.2 m³, 1.6 km, 0.2%
FR-3	Toxic dispersion	40.4 m³, 1.6 km, 15.3%	100.9 m³, 1.6 km, 4.6%	1008.8 m³, 1.6 km, 1.8%	2017.6 m³, 1.6 km, 0.2%
FR-4	Toxic dispersion	60.5 m³, 2.4 km, 15.2%	151.3 m³, 2.4 km, 4.4%	1512.6 m³, 2.4 km, 1.7%	3025.2 m³, 2.4 km, 0.2%
FR-5	Toxic dispersion	50.4 m³, 2.4 km, 15.2%	126 m³, 2.4 km, 4.4%	1260 m³, 2.4 km, 1.8%	2521 m³, 2.4 km, 0.2%
FR-6	Toxic dispersion	40.3 m³, 2.4 km, 15.3%	100.8 m³, 2.4 km, 1.8%	1008.4 m³, 2.4 km, 1.8%	2016.8 m³, 2.4 km, 0.2%
S-1	Toxic dispersion	No release	15.6 m³, 0.6 km, 3.4%	155.5 m³, 1.9 km, 0.3%	777.6 m³, 8.7 km, 0.01%
S-2	Toxic dispersion	No release	48.0 m³, 1.0 km, 3.5%	480.3 m³, 2.9 km, 0.3%	2401.6 m³, 13.5 km, 0.01%
T-1	Pool fire	16.9 m³, 33.5 m, 0.07%	42.4 m³, 33.5 m, 0.2%	423.6 m³, 33.5 m, 0.2%	847.2 m³, 88.0 m, 0.02%
T-2	Pool fire	8.5 m³, 33.5 m, 0.07%	21.2 m³, 33.5 m, 0.2%	212.0 m³, 33.5 m, 0.2%	424.0 m³, 33.5 m, 0.02%
T-3	Pool fire	8.1 m³, 27.9 m, 0.07%	20.2 m³, 27.9 m, 0.2%	201.6 m³, 27.9 m, 0.2%	403.2 m³, 27.9 m, 0.02%
T-4	Pool fire	8.7 m³, 23.1 m, 0.07%	21.7 m³, 23.1 m, 0.2%	217.4 m³, 23.1 m, 0.2%	434.8 m³, 23.1 m, 0.02%
T-5	Pool fire	35.0 m³, 40.4 m, 0.1%	87.4 m³, 40.4 m, 0.04%	874.4 m³, 40.4 m, 0.02%	1748.8 m³, 151.2 m, 0.002%
T-6	Pool fire	13.0 m³, 26.6 m, 0.07%	32.4 m³, 26.6 m, 0.2%	324 m³, 26.6 m, 0.2%	648.0 m³, 26.6 m, 0.02%

The table shows tank volume involved in the accidents, end-point distance, and event probability.

FIGURE 10.7 Natech Impact Areas for Scenario Earthquake Impact on All Storage Tanks (Map Data ©2016 Google)

When subjected to the seismic forces of the Istanbul earthquake scenario, RAPID-N predicts releases from all tanks. Consequently, the maximum end-point distances increase significantly to a maximum value of 13.5 km with the atmospheric dispersion of ammonia for the worst-case scenario of complete release of content from tank S-2. Fig. 10.7 shows that in this case also urban areas at the shore of Izmit Bay opposite the case-study installation could be at risk of exposure if the wind blows in this direction. Fig. 10.8 provides a close-up of the expected Natech end-point distances in the vicinity of the installation. The end-point circles are concentrated in an area of 2.4 km around the tank farm, making this the most critical region for suffering toxic effects.

For tanks with similar dimensions and structural characteristics (e.g., tanks FR-1 to FR-6) RAPID-N calculates similar earthquake damage probabilities and related Natech event probabilities. The end-point distances are also similar and differ only due to the substance type (e.g., acrylonitrile vs. vinyl acetate) because in most of the cases the containment dikes were large enough to hold the released material for all damage states and hence they reduced the pool surface area, which restricted evaporation. The effect of H/D ratio on the damage probability can be clearly seen in Table 10.4 between floating-roof (H/D < 0.7) and fixed-roof tanks (H/D ≥ 0.7, except T-5). Increasing the H/D ratio shifts the higher damage probabilities from lower damage states (DS1–DS2) to higher damage states (DS3–DS4), although the difference between the estimated probability values is less than one order of magnitude. Nevertheless, in this case study the Natech probabilities for fixed-roof tanks

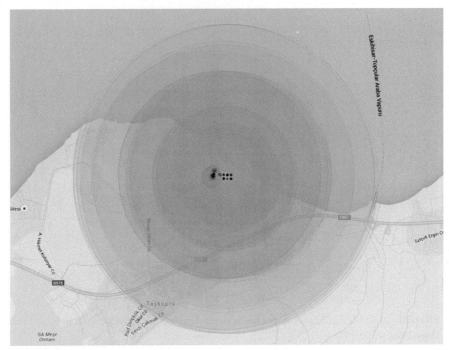

FIGURE 10.8 Close-Up of End-Point Distances in the Vicinity of the Facility (Map Data ©2016 Google)

were found to be much smaller than those for the other atmospheric tanks. Since the fixed-roof tanks contain flammable substances that ignite, the release probability needs to be multiplied with the conditional ignition probability, which decreases the final consequence probability for these tanks.

For the conditions of the Istanbul earthquake, RAPID-N predicts releases from all six fixed-roof tanks containing methanol and acetic acid. As already mentioned in Section 10.3.2, the end-point distances for heat radiation from the pool fires originating from these tanks envelop other tank locations. The potential for domino effects is therefore evident and should be considered for a more accurate assessment of the Natech risk from these tanks. It should also be noted that although an increased minimum pool depth is utilized to better simulate pool dimensions in case of dike overflows, adjoining dikes as in the case of this facility may result in smaller pool dimensions if the overflow runs to other dikes. A more detailed analysis is suggested for such scenarios.

10.4 CONCLUSIONS

Using the ground-motion parameters predicted for the Istanbul earthquake scenario as input, a Natech risk analysis of a hazardous installation in Izmit Bay was carried out using the RAPID-N framework. The case study showed the capability of RAPID-N to analyze and map the impact of earthquakes on a single plant unit but also on multiple plant units containing different types of substances simultaneously. Similarly, the tool is also capable of analyzing multiple installations with multiple plant units concurrently, which is useful for regional Natech risk analysis and mapping.

While a number of simplifying assumptions were made to compensate for the lack of detailed data on the industrial plant and the substance hazard, the results of the study indicate that possibly major Natech accidents are to be expected in case of the predicted Istanbul earthquake. These results are in agreement with historical data of Natech damage in the area due to the 1999 Kocaeli earthquake (Girgin, 2011). Due to limited data availability and the assumptions made, the results may not reflect the actual Natech risk of the facility. Therefore, they must be regarded as indicative and should not be used for decision making without careful validation. In collaboration with AFAD, other case studies with RAPID-N will be performed in Turkey that will include more detailed data collection and validation. Several test regions are currently under discussion. More detailed risk-state scenarios for different types of plant units under earthquake loading will be identified via the analysis of historical accident data and implemented in RAPID-N for a more comprehensive Natech risk analysis.

Acknowledgments

The authors acknowledge the Turkish Prime Ministry Disaster and Emergency Management Presidency (AFAD) for providing information on the Istanbul earthquake scenario.

References

AIHA, 1988. Emergency Response Planning Guidelines (ERPG) & Workplace Environmental Exposure Levels (WEEL) Handbook. American Industrial Hygiene Association, Falls Church VA.

Ambraseys, N.N., 1970. Some characteristic features of the Anatolian fault zone. Tectonophysics 9, 143–165.

API, 1990. API RP 510 Pressure Vessel Inspection Code: Maintenance, Inspection, Rating, Repair, Alteration. American Petroleum Institute, Washington DC.

Barka, A., 1992. The North Anatolian fault zone. Annales Tectonicae 6, 164–195.

Boore, K.M., Joyner, W.B., Fumal, T.E., 1997. Equations for estimating horizontal response spectra and peak acceleration from Western North American earthquakes: a summary of recent work. Seismol. Res. Lett. 68 (1), 128–153.

Campbell, K.W., 1997. Empirical near-source attenuation relationships for horizontal and vertical components of peak ground acceleration, peak ground velocity, and pseudo-absolute acceleration response spectra. Seismol. Res. Lett. 68 (1), 154–179.

Cox, A.W., Lees, F.P., Ang, M.L., 1990. Classification of hazardous locations, Institution of Chemical Engineers, Rugby.

Cozzani, V., Gubinelli, G., Salzano, E., 2006. Escalation thresholds in the assessment of domino accidental events. J. Hazard. Mater. 129 (1–2), 1–21.

Durukal, E., Erdik, M., 2008. Physical and economic losses sustained by the industry in the 1999 Kocaeli, Turkey earthquake. Nat. Hazards 46 (2), 153–178.

EU, 2008. Regulation on classification, labelling and packaging of substances and mixtures, amending and repealing Directives 67/548/EEC and 1999/45/EC, and amending Regulation (EC) No 1907/2006. Off. J. Eur. Union 353, 1–1355.

FEMA, 2010. Multi-Hazard Loss Estimation Methodology, Earthquake Model, HAZUS® MH MR5, Technical Manual. US Federal Emergency Management Agency, Washington, DC.

Girgin, S., 2011. The natech events during the 17 August 1999 Kocaeli earthquake: aftermath and lessons learned. Nat. Hazards Earth Syst. Sci. 11, 1129–1140.

Grünthal, G. (Ed.), 1998. European Macroseismic Scale 1998 (EMS-98). European Center for Geodynamics and Seismology, Luxembourg.

Hubert-Ferrari, A., Barka, A.A., Jacquess, E., Nalbant, S.S., Meyer, B., Armijo, R., Tapponier, P., King, G.C.P., 2000. Seismic hazard in the Sea of Marmara following the Izmit earthquake. Nature 404, 269–272.

IARC, 1999. IARC monographs on the evaluation of carcinogenic risks to humans. Re-Evaluation of Some Organic Chemicals, Hydrazine and Hydrogen Peroxide, vol. 71, IARC Working Group on the Evaluation of Carcinogenic Risks to Humans, Lyon, France.

JICA, 2002. The study on a disaster prevention/mitigation basic plan in Istanbul including seismic microzonation in the Republic of Turkey, Final Report. Japan International Cooperation Agency, Istanbul.

McClusky, S., et al., 2000. Global Positioning System constraints on plate kinematics and dynamics in the eastern Mediterranean and Caucasus. J. Geophys. Res. 105, 5695–5719.

Moschonas, C., Karakostas, C., Lekidis, V., Papadopoulos, S., 2014. Investigation of seismic vulnerability of industrial pressure vessels. Proceedings of the Second European Conference on Earthquake Engineering and Seismology, 25–29 August 2014, Istanbul.

O'Rourke, M.J., So, P., 2000. Seismic fragility curves for on-grade steel tanks. Earthquake Spectra 16 (2), 801–815.

Parsons, T., 2004. Recalculated probability of M ≥ 7 earthquakes beneath the Sea of Marmara, Turkey. J. Geophys. Res. 109, 1–21.

Parsons, T., Toda, S., Stein, R.S., Barka, A., Dieterich, J.H., 2000. Heightened odds of large earthquake near Istanbul: an interaction-based probability calculation. Science 288, 661–665.

Rahnama, M., Morrow, G., 2000. Performance of industrial facilities in the August 17, 1999 Izmit Earthquake. Proceedings of the Twelfth World Conference on Earthquake Engineering, 2851, Auckland, New Zealand.

Stein, R.S., Barka, A., Dieterich, J.H., 1997. Progressive failure on North Anatolian Fault since 1939 by earthquake stress triggering. Geophys. J. Int. 128, 594–604.

Steinberg, L.J., Cruz, A.M., 2004. When natural and technological disasters collide: lessons from the Turkey earthquake of August 17, 1999. Nat. Hazards Rev. 5 (3), 121–130.

Straub, C., Kahle, H.G., Schindler, C., 1997. GPS and geologic estimates of the tectonic activity in the Marmara Sea region, NW Anatolia. J. Geophys. Res. 102 (27), 587–601.

Suzuki, K., 2002. Report on damage to industrial facilities in the 1999 Kocaeli Earthquake, Turkey. J. Earthquake Eng. 6 (2), 275–296.

TNO, 1996. Methods for calculation of physical effects due to releases of hazardous materials (liquids and gases), CPR 14E. The Netherlands Organization of Applied Scientific Research (TNO), The Hague.

U.S. EPA, 1999. Risk Management Program Guidance for Offsite Consequence Analysis. U.S. Environmental Protection Agency, Chemical Emergency Preparedness and Prevention Office, USA.

Wald, D.J., Quitoriano, V., Heaton, T.H., Kanamori, H., Scrivner, C.W., Worden, C.B., 1999. TriNet "ShakeMaps": rapid generation of peak ground motion and intensity maps for earthquakes in Southern California. Earthquake Spectra 15 (3), 537–555.

Case-Study Application II: ARIPAR-GIS

G. Antonioni, A. Necci, G. Spadoni, V. Cozzani
Department of Civil, Chemical, Environmental, and Materials Engineering,
University of Bologna, Bologna, Italy

This chapter demonstrates the detailed quantitative analysis of Natech risk by applying the ARIPAR-GIS software introduced in Chapter 9 to two case studies. Case study 1 analyzes earthquake-induced Natech risks at a hazardous facility and case study 2 studies the impact of floods. Individual and societal risk were calculated and compared to the risk levels obtained without considering Natech scenarios. The results confirm the significant influence that Natech risks may have on the overall risk at a hazardous installation.

11.1 INTRODUCTION

This chapter discusses the analysis of two Natech case studies to understand the importance of the assessment of industrial risk associated with natural events. The first case study is concerned with the analysis of Natech accidents triggered by earthquakes, the second one with accidents triggered by floods. The general approach to quantitative risk analysis outlined in Chapter 7 and the specific procedure summarized in Chapter 9 (Antonioni et al., 2009; Cozzani et al., 2014) were applied to the case study calculations. The use of a software tool was a necessary step in the process and the Natech module of the ARIPAR-GIS software, introduced in Chapter 9, was applied (Antonioni et al., 2007). The two case studies were based on layouts and process equipment data derived from those of existing chemical, and oil and gas facilities.

11.2 CASE STUDY 1: NATECH SCENARIOS TRIGGERED BY EARTHQUAKES

The first simplified case study is dedicated to the assessment of the contribution of earthquake-induced Natech scenarios to the overall individual and societal risk. The layout used for the case study is presented in Fig. 11.1. As shown in this figure, eight atmospheric tanks were considered in the analysis. For the sake of simplicity, all the tanks were assumed to have the same volume (3500 m³) and the same content (2000 t of ethanol).

Natech Risk Assessment and Management. http://dx.doi.org/10.1016/B978-0-12-803807-9.00011-5

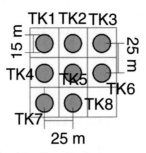

FIGURE 11.1 Layout Considered for Case Study 1

Adapted from Antonioni et al. (2007).

In order to simplify the analysis, a single scenario was associated with each equipment item and was considered as the only possible primary and/or secondary event. The scenarios were defined on the basis of credible accidents involving the equipment items considered, also following the suggestions given by the *Purple Book* (Uijt de Haag and Ale, 1999). The case study mainly aimed to analyze events triggered by earthquakes, thus only severe scenarios were taken into account. The instantaneous release of the entire content of the tank was considered as the reference scenario. A frequency of 3.1×10^{-7} events/year was assumed for internal failures (conventional failures not induced by earthquakes). With respect to earthquakes, a reference event with a PGA of 0.224 g and a return period of 475 years was assumed. The damage probability was calculated using the probit models listed in Table 9.2, and it was equal to 0.026 for each of the eight tanks in the case study. A total of 255 (2^8–1) possible Natech accident combinations were assessed. The final outcome of the scenario was pool fire of the entire catch-basin area.

The conventional consequence-analysis models described in the *Yellow Book* (van den Bosch and Weterings, 2005) were used for the assessment. For calculating the vulnerability of the surrounding population, the probit models in Table 11.1 were used. A fictitious value of 5 persons/ha and a homogeneous density were assumed for the population distribution.

Table 11.1 Probit Functions for Human Vulnerability Used in the Case Studies (Mannan, 2005; van den Bosch et al., 1992)

Scenario	Target	Probit Equation	Dose, D	Dose Units
Radiation	Human	$Y = -14.9 + 2.56 \ln(D)$	$I^{1.33} \cdot t_e$	I: kW/m² t_e: s
Overpressure	Human	$Y = 1.47 + 1.37 \ln(D)$	p_s	p_s: psig
Toxic release: NH_3	Human	$Y = -9.82 + 0.71 \ln(D)$	$C^2 \cdot t_e$	C: ppm t_e: min

Y, *probit value*; I, *radiation intensity*; p_s, *peak static overpressure*; C, *toxic concentration*, t_e, *exposure time*.

The results of the quantitative risk assessment using ARIPAR-GIS are shown in Fig. 11.2, both in terms of local specific individual risk (LSIR) (A) and of societal risk (B). As is clearly visible from this figure, the earthquake causes a significant increase in individual and societal risk levels, in particular for nonanchored tanks but also for those that are anchored, albeit at a somewhat lesser level. Two effects are evident from Fig. 11.2 in accordance with theoretical considerations:

1. An increase in the values of the frequency, F, corresponding to the reference scenarios chosen for each unit: this is caused by the increase in the overall frequency of the reference scenarios due to the possibility that the equipment may fail also due to an earthquake.
2. An increase in the maximum value of expected fatalities, N, caused by the assumption that seismic events may trigger scenarios simultaneously involving more than one unit. This assumption is never introduced in conventional QRA unless domino events are considered (Cozzani et al., 2005; Reniers and Cozzani, 2013). Clearly, assuming that several reference scenarios may take place at the same time results in overall events having a higher overall value of expected fatalities than that of the single reference scenarios.

This case study showed that the consideration of accident scenarios due to earthquakes adds important contributions to the overall values of these risk indices.

11.3 CASE STUDY 2: NATECH SCENARIOS TRIGGERED BY FLOODS

11.3.1 Layout and Vessel Features

In order to demonstrate the application of the methodology and to understand the importance of considering flood-induced Natech scenarios, a case-study QRA with ARIPAR-GIS was carried out. The layout of the industrial facility selected for the study is shown in Fig. 11.3. It should be noted that in the layout both atmospheric and pressurized tanks are present.

Table 11.2 lists the features of the vessels considered and their inventories of hazardous substances. Both horizontal and vertical tanks were included in the analysis. All the horizontal vessels were assumed to be supported on a concrete base (0.25 m above ground level) that may provide protection from floods with low water depths.

11.3.2 Workers and Surrounding Population

The industrial facility considered is organized for continuous operation 24 h/day. Thus, inside the site, a constant presence of 100 workers was considered. The workers were assumed to be evenly distributed in the plant area. A 50% probability of being present outdoors was considered.

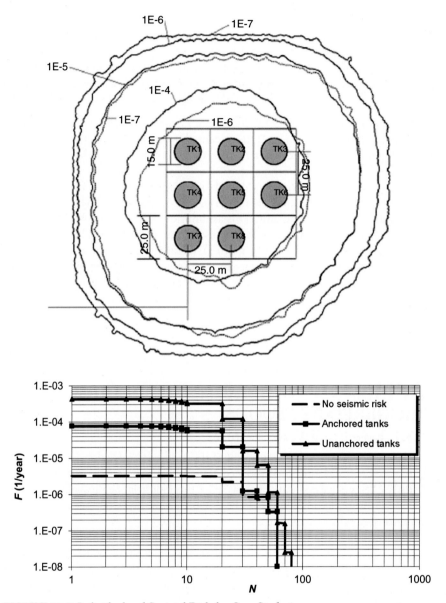

FIGURE 11.2 Individual and Societal Risk for Case Study 1

(A) LSIR contours (events/year), (B) societal risk with (*solid lines*) and without (*dashed lines*) the consideration of possible accident scenarios due to earthquake impact. The impact of anchoring tanks as a safety measure is also shown.

Adapted from Antonioni et al. (2007).

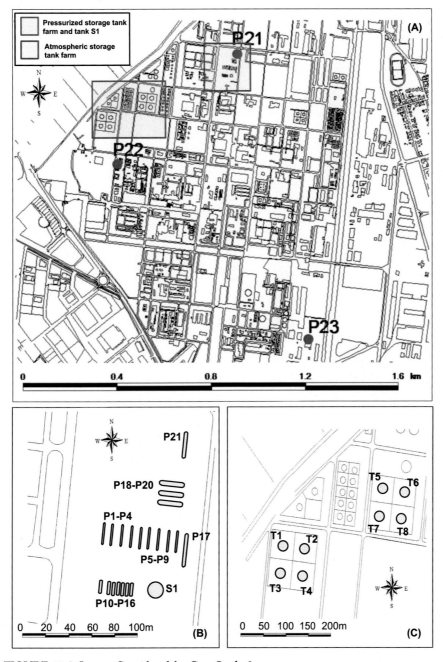

FIGURE 11.3 Layout Considered for Case Study 2

(A) Overview of the industrial area, position of the tank farms, and location of pressurized tanks P21–P23; (B) pressurized tank farm and storage tank S1; (C) atmospheric tank farm.

Table 11.2 *Main Features of the Vessels Considered in Case Study 2 With Ambient Temperature 293 K*

Vessel Features	Pressurized Vessels					Atmospheric Vessels		
	P1–P9	P10–P16	P17	P18–P20	P21–P23	S1	T1–T4	T5–T8
Nominal capacity (m^3)	50	30	115	150	100	3179	6511	6511
Diameter (m)	2.7	2.4	2.75	3.2	2.8	15	24	24
Length^a/height^b (m)	10	6.5	20.1	19.4	18	18	14.4	14.4
Shell thickness (mm)	23	21	24	27	24	12.5	12.5	12.5
Vessel tare weight (metric ton)	12.3	5.9	29.2	36.1	26.2	110	165	165
Saddle parameter (m)	1.48	1.38	1.58	1.78	1.58	—	—	—
Filling level	90%	90%	90%	90%	90%	75%	75%	75%
Substance contained	Propylene	Propane	LPG^c	Ammonia	Chlorine	Organic solvent	Gasoline	Benzene
Physical state	Liquefied gas	Liquefied gas	Liquefied gas	Liquefied gas	Liquefied gas	Liquid	Liquid	Liquid
Pressure (bar)	8	8.5	2	8.5	6.7	1.05	1.05	1.05
Liquid density (kg/m^3)	615	450	550	600	1400	650	750	877
Vapor density (kg/m^3)	13.8	15.4	4.8	4.9	19.3	0.97^d	0.97^d	0.97^d
Inventory (metric ton)	32	12	59	84	140	1550	3656	4275

[a]Horizontal vessel.
[b]Vertical vessel.
[c]Assumed as pure butane.
[d]Average density of the purge gas (e.g., nitrogen blanketing), not relevant for model application
Adapted from Antomioni et al. (2015).

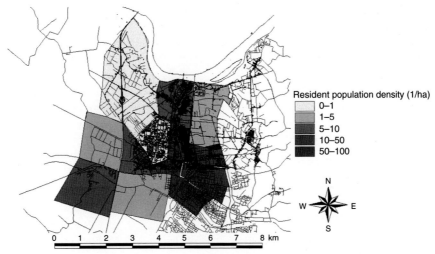

FIGURE 11.4 Density of the Resident Population in the Surroundings of the Industrial Area

Adapted from Antonioni et al. (2015).

Census data were used as the basis for societal-risk calculation. Fig. 11.4 shows the distribution considered for the resident population. Daily averages for presence probability of resident population were considered (Uijt de Haag and Ale, 1999; Bonvicini et al., 2012).

11.3.3 Flood Scenarios

The flood scenarios considered in case study 2 are summarized in Table 11.3. Four reference scenarios were selected in order to consider different types of flood waves. In particular, extremely severe conditions were assumed for the first two reference scenarios (Scenario 1 and Scenario 2). In the first scenario, high-depth flooding with limited speed was taken into account. In contrast, in the second scenario a "flash-flood," with high speed but low water depth was assumed. Both conditions are associated with low frequency values (see Table 11.3). The other reference scenarios were associated with lower-severity flood conditions having a lower return time, that is, they occur more frequently. The defined reference flood scenarios allow the assessment of the impact of different types of floods, with different damage potential and expected frequency.

11.3.4 Individual and Societal Risk Calculated for Conventional Scenarios

In order to understand the importance of Natech scenarios triggered by floods, as in the previous case study a QRA of "conventional" scenarios, due to internal failures at the installation, was performed to obtain reference values for individual and societal

Table 11.3 Flood Reference Scenarios Defined for Case Study 2

Flood Conditions	Return Time (Year)	Flood Frequency (1/Year)	Flood Depth (m)	Flood Velocity (m/s)
Scenario 1	500	2.0×10^{-3}	2.00	0.5
Scenario 2	500	2.0×10^{-3}	0.50	2
Scenario 3	200	5.0×10^{-3}	1.15	0.75
Scenario 4	30	3.33×10^{-2}	0.75	0.5

Adapted from Antonioni et al. (2015).

Table 11.4 End-Point Scenarios Considered for the QRA of Internal Release Events at a Hazardous Facility

Tank ID	Loss of Containment Event	Final Outcome	Frequency (1/Year)
T1–T8 and S1	Instantaneous release into the catch basin of the total inventory	Pool fire	4.5×10^{-6}
P1–P17	Release in 10 min of the total inventory	Flash fire of propane, propylene, or LPG	4.5×10^{-7}
P18–P23	Release in 10 min of the total inventory	Toxic cloud of ammonia or chlorine	5.0×10^{-7}

Adapted from Antonioni et al. (2015).

risk. The expected frequency of the top events was defined according to the *Purple Book* (Uijt de Haag and Ale, 1999). Table 11.4 shows the end-point frequencies of each scenario considered for the risk sources analyzed. The consequences of the end-point events listed in the table were assessed using literature models (Mannan, 2005; CCPS, 2000; van Den Bosch and Weterings, 2005). The physical effects calculated were then implemented in the ARIPAR-GIS software.

Fig. 11.5 shows the individual risk calculated for the conventional scenarios included in the case study. The risk contour at the threshold value of 10^{-6} 1/year lies inside the industrial area, while only lower individual risk levels are present in the residential areas. A Potential Life Loss (PLL) of 8.84 fatalities per thousand years was calculated. These results are the baseline values for the comparison with the risk levels calculated for the accident scenarios triggered by floods.

11.3.5 Individual and Societal Risk Including Flood-Induced Scenarios

The methodology for the quantitative assessment of Natech scenarios in QRA studies discussed in Chapter 9 was applied to the analysis of the four reference flood events considered. The equipment vulnerability models in Tables 9.3 and 9.4, derived from

FIGURE 11.5 Individual-Risk Contours (1/Year) Calculated for Accident Scenarios Deriving From Conventional Hazmat Release Events due to Internal Failures

the studies of Landucci et al. (2012, 2014) were applied in the quantitative calculations of vessel failure probability (cf. Chapter 9). For atmospheric tanks, the approach summarized in Table 9.3 was applied, thus determining the critical filling level (CFL) as a function of the stored substance, tank geometry, and flood conditions. As shown in the table, for all the flood conditions considered, the failure of the atmospheric tanks resulted credible. The failure probabilities given the flood ranged from 1%–5% to 15%–20%, respectively, for the low- and high-severity flood scenarios.

In the case of pressurized horizontal vessels, completely different results were obtained. For these vessels, the fragility model described in Table 9.4 was used. Even in the presence of a high flood velocity, the "critical velocity" is not exceeded for any of the flood reference scenarios listed in Table 11.3 (Landucci et al., 2014). Hence, the tank failure probability was evaluated only according to the estimated CFL, resulting in high values for flood scenario 1 (up to 100% failure probability), while for the other flood scenarios low values of failure probability were obtained (down to 0% failure probability in flood scenario 4). This was due to the fact that the tanks were considered anchored to concrete supports, which limited the lift forces associated with the flood.

On the basis of the calculated equipment failure probabilities, the application of the detailed procedure for Natech QRA discussed in Chapter 9 allowed the calculation of the frequencies and probabilities of scenarios involving the simultaneous damage of more than one equipment item. Fig. 11.6 shows the results of the quantitative risk analysis with ARIPAR-GIS in the presence of Natech scenarios in terms of LSIR

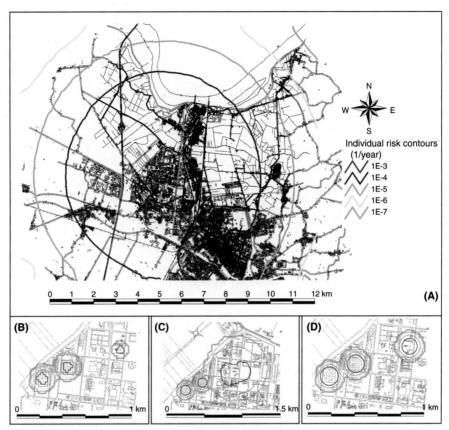

FIGURE 11.6 Individual-Risk Contours (1/Year) due to Flood-Triggered Accidents

(A) Flood scenario 1, (B) flood scenario 2, (C) flood scenario 3, (D) flood scenario 4.

From Antonioni et al. (2015).

contours. Flood scenario 1, in which a high depth of flood water is assumed, resulted in the most critical flood event, leading to the highest number of damaged vessels and accident scenarios. On the basis of the failure frequencies estimated from the equipment vulnerability models, about 30,000 scenarios (over a total number of $2^{32}-1$ possible combinations) resulted in a frequency above the cut-off value of 10^{-10} 1/year, but only 11,000 of them contributed significantly to the overall risk.

Flood scenario 2, which can be considered as a flash-flood due to its high water speed and limited depth, does not contribute significantly to the risk indices because several equipment items (in particular pressurized vessels) are mounted on concrete supports having a height comparable to the maximum flood-water depth considered. Thus, for scenario 2, damage was predicted only for the nine atmospheric tanks, and "only" $2^9-1 = 511$ scenarios were considered. Among those, the frequencies of merely 185 were above the cut-off value. Flood scenarios 3 and 4, which

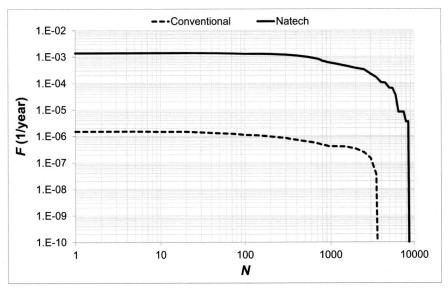

FIGURE 11.7 Societal Risk Evaluated for Conventional Scenarios due to Internal Failures and for Flood Scenario 1

Adapted from Antonioni et al. (2015).

concern less severe but more frequent flood events, also resulted in only a low contribution to the overall risk.

Fig. 11.7 compares the *F–N* curve for societal risk that includes flood Natech scenarios with the one that considers only conventional accident scenarios. The societal risk was calculated assuming a specific population distribution around the industrial area. In the case of flood-induced Natech scenarios, the conservative assumption that the population distribution will not change during a flood was introduced in the calculations.

The *F–N* curve that considers Natech events is mainly influenced by the increase in the frequency of loss of containment from chlorine tanks (P21–P23) due to flooding. Only the toxic cloud dispersion of chlorine from the rupture of pressurized vessels P21, P22, and P23 resulted in physical effects sufficiently high to cause harm in the areas where the resident population was present. Since only the reference flood event considered for flood scenario 1 was sufficiently severe to affect these vessels, this was the only flood event that caused changes in the overall societal risk value. The *F–N* curve that includes Natech scenarios also shows some additional steps at high *N* values (*N* > 4000) due to the presence of "combined scenarios," where the impact of the simultaneous failure of and toxic releases from some, or all, of the chlorine tanks due to the flood event is considered.

The potential life loss (PLL) increases from 8.84×10^{-3} fatalities/year to 14.8 fatalities/year when including Natech events due to their higher frequency and severity if compared to conventional accidents.

11.4 RESULTS OF THE CASE-STUDY ANALYSES

The results of the case studies demonstrate that Natech scenarios can have a high impact on the risk profile of industrial facilities storing and processing hazardous materials. These significant risk levels are mostly attributable to the rather high calculated values of vessel failure frequencies due to the high frequency of severe natural events estimated on the basis of the return time. This issue is also pointed out in several studies (Landucci et al., 2012, 2014) in which Natech frequencies were compared with baseline frequency values used for component failure in "conventional" QRA. In fact, in areas exposed to natural hazards, the frequencies of such events may reach values that are orders of magnitude higher with respect to those related to component failures due to internal causes (e.g., mechanical failure, corrosion, erosion, rupture induced by vibrations, and so on).

The analyses also showed that atmospheric tanks are the most vulnerable tank category, with possible failures even in the presence of low-severity natural events. Therefore, the results of quantitative risk analysis allow the determination of the most vulnerable equipment items and the units that may lead to most severe accident scenarios. The case studies exemplified that a risk-based approach supports the identification of the most critical items in the plant with respect to several types of natural hazards that may affect a given geographical area. Hence, the results may be used to support risk-informed decision making concerning the protection of industrial sites where hazardous substances are present.

Finally, it is also worth noting that the extremely high level of societal risk in Fig. 11.7 is also due to the assumption that no change in time of the surrounding population's presence was considered. On the one hand, in the case of earthquakes, flash floods, or floods caused by the sudden rupture of river levees, no time is usually available for warning and evacuation, and the obtained overall risk may therefore be considered realistic. On the other hand, in the case of long rivers for which forecasting models are available, floods due to heavy rains or other adverse weather conditions may be anticipated by several hours or even days. Consequently, for slow-onset flood scenarios there may be time for early warning and evacuation of the population and workers that may significantly change the number of persons exposed to the effects of the flood and of flood-induced Natech scenarios. The final consequences of flood-induced Natech scenarios may therefore be different if the evacuation of the population is considered.

References

Antonioni, G., Spadoni, G., Cozzani, V., 2007. A methodology for the quantitative risk assessment of major accidents triggered by seismic events. J. Hazard. Mater. 147, 48.

Antonioni, G., Bonvicini, S., Spadoni, G., Cozzani, V., 2009. Development of a framework for the risk assessment of Na-tech accidental events. Reliab. Eng. Syst. Saf. 94, 1442.

Antonioni, G., Landucci, G., Necci, A., Gheorghiu, D., Cozzani, V., 2015. Quantitative assessment of risk due to Natech scenarios caused by floods. Reliab. Eng. Syst. Saf. 142, 334.

Bonvicini, S., Ganapini, S., Spadoni, G., Cozzani, V., 2012. The description of population vulnerability in Quantitative Risk Analysis. Risk Anal. 32, 1576.

CCPS, 2000. Guidelines for Chemical Process Quantitative Risk Analysis, second ed. American Institute of Chemical Engineers, Center for Chemical Process Safety, New York.

Cozzani, V., Gubinelli, G., Antonioni, G., Spadoni, G., Zanelli, S., 2005. The assessment of risk caused by domino effect in quantitative area risk analysis. J. Hazard. Mater. 127 (1–3), 14.

Cozzani, V., Antonioni, G., Landucci, G., Tugnoli, A., Bonvicini, S., Spadoni, G., 2014. Quantitative assessment of domino and Natech scenarios in complex industrial areas. J. Loss Prev. Process Ind. 28, 10.

Landucci, G., Antonioni, G., Tugnoli, A., Cozzani, V., 2012. Release of hazardous substances in flood events: damage model for atmospheric storage tanks. Reliab. Eng. Syst. Saf. 106, 200.

Landucci, G., Necci, A., Antonioni, G., Tugnoli, A., Cozzani, V., 2014. Release of hazardous substances in flood events: damage model for horizontal cylindrical vessels. Reliab. Eng. Syst. Saf. 132, 125.

Mannan, S., 2005. Lees' Loss Prevention in the Process Industries, third ed. Elsevier, Oxford.

Reniers, G., Cozzani, V., 2013. Domino Effects in the Process Industries—Modeling, Prevention and Managing. Elsevier, Amsterdam.

Uijt de Haag, P.A.M., Ale, B.J.M., 1999. Guidelines for Quantitative Risk Assessment. Purple Book. Committee for the Prevention of Disasters, The Hague.

van Den Bosch, C.J.H., Weterings, R.A.P.M., 2005. Methods for the calculation of physical effects. Yellow Book. third ed. Committee for the Prevention of Disasters, The Hague.

van Den Bosch, C.H.J., Merx, W.P.M., Jansen, C.M.A., De Weger, D., Reuzel, P.G.J., Leeuwen, D.V., Blom-Bruggerman, J.M., 1992. Methods for the calculation of possible damage. Green Book. Committee for the Prevention of Disasters, The Hague.

Case Study Application III: RISKCURVES

A. Basco*, I. Raben**, J. Reinders**, E. Salzano†
*AMRA, Analysis and Monitoring of Environmental Risk, Naples, Italy;
**TNO-The Netherlands Organization for Applied Scientific Research,
Utrecht, The Netherlands; †Department of Civil, Chemical, Environmental,
and Materials Engineering, University of Bologna, Bologna, Italy

This chapter discusses the application of TNO's software packages RISKCURVES and EFFECTS in the quantitative Natech risk analysis of a refinery located on the Mediterranean coast. In addition to conventional industrial accident scenarios that could lead to a release of dangerous substances, the impact of two natural hazards (earthquake and tsunami) on the overall risk level was also considered and the priority risk contributors identified.

12.1 INTRODUCTION

Public perception of the disaster potential derived from the interaction of natural hazards with industrial installations has strongly increased in the last decades (Krausmann et al., 2011; Salzano et al., 2013). For effective Natech risk reduction to take place, the risk first needs to be identified and analyzed. Hence, there is a strong need for Natech risk-analysis methodologies and tools. In this study, a quantitative analysis of Natech risks of a refinery at a Mediterranean coast was carried out using RISKCURVES and EFFECTS, exploiting the fact that computerized tools have been used for decades for the assessment of conventional industrial risks. Due to the refinery's location, the impact of earthquakes and tsunamis was analyzed in addition to risks from accident causes other than natural hazards.

The analysis was performed within the framework of the EU project STREST (2016) which aimed to develop harmonized stress tests for critical infrastructures against natural hazards and which followed the stress tests performed for European Union nuclear power plants to review their response to extreme situations.

12.2 METHODOLOGY

Quantitative risk analysis (QRA) is a complex and time-consuming task, as depicted in Fig. 12.1, which shows the steps and relevant aspects of a QRA. Selecting relevant equipment, choosing which accident scenarios are applicable, determining

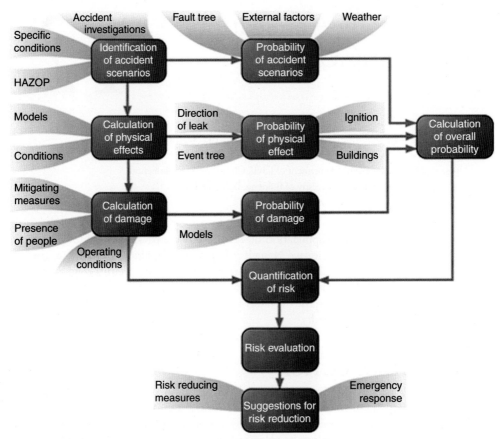

FIGURE 12.1 Steps for Standard Industrial Quantitative Risk Assessment

their frequencies, and modeling the physical effects are just a few of the hurdles a user finds on his way when performing a QRA. These issues clearly affect the overall risk figures as several arbitrary choices and simplifications are necessarily introduced in the process. For the sake of comparability of studies, standard methods were developed in the Netherlands, to a large extent by TNO, through reference handbooks which are called the *Yellow Book*, *Green Book*, *Purple Book*, and *Red Book* (VROM, 2005a–d). The "colored" books are public and may be considered de facto as a standard reference for any risk analysis. The QRA presented in this chapter was performed using TNO's software packages RISKCURVES for quantitative risk assessment and EFFECTS for safety and hazard analysis (TNO, 2015), both of which are based on the "colored" books.

For any defined accident scenario, the physical effects and consequences need to be assessed. These are influenced by the release rate, weather conditions, and calculation models/methods, etc. Physical effects evaluated are toxic concentrations, heat radiation (pool fire, flash fire, torch fire, fire ball), and overpressure. The *Yellow Book* contains models for the calculation of physical effects. Vulnerability models relating heat radiation, overpressure, and toxic concentrations to lethal consequences are provided in the *Green Book* and were implemented in the same software. The probability of scenarios, physical effects, and damage is influenced by many parameters. It can be assessed by using fault trees, event trees, case histories, or following the *Red* and the *Purple Book*. Risk is finally determined from two attributes: the adverse effects (consequences) of an accident and the frequency with which these consequences occur. In the QRA study presented in this chapter, two types of risks are considered based on Dutch legislation (Ministry of Infrastructure and Environment, 2010): locational risk (LR) and societal risk.

The first is defined as the frequency per year that a hypothetical person will be lethally affected by the consequences of possible accidents during an activity involving hazardous materials, for example, a chemical plant or transport activities. This risk indicator is a function of the distance between the exposed person and the activity, regardless of whether people are actually living in the area, or at the specified location. LR is presented in contours on maps of the surroundings; the contours connect locations of equal LR. The LR is sometimes also referred to as the individual risk (Section 7.2). Societal risk is defined as the cumulative frequency that a minimum number of people will simultaneously be killed due to possible accidents during an activity with hazardous materials. Here, the actual presence of people in the surroundings is taken into account. The societal risk for transport activities is calculated per single kilometer. Societal risk is also called "group risk."

According to studies by several authors, the overall methodology proposed in Fig. 12.1 can be applied when considering natural hazards as the triggering event for the structural failure of industrial equipment, provided that the vulnerability of the equipment is evaluated in terms of loss of containment of hazardous materials (Fabbrocino et al., 2005; Campedel et al., 2008; Salzano et al., 2009).

12.3 DESCRIPTION OF THE CASE STUDY

A refinery on the Mediterranean coast was selected for this study. In the considered installation many storage tanks are present, containing a large variety of hydrocarbons, such as LPG, gasoline, gasoil, crude oil, and atmospheric and vacuum residues. The capacities of the tanks vary from about 100 m^3 (fuel oil, gasoil, gasoline, kerosene) to 160,000 m^3 (crude oil). All tanks are located in catch basins (bunds) with concrete surfaces. Solely the LPG is stored in pressurized spheres, all others in single containment tanks.

In order to evaluate the societal risk, the (actual) presence of persons in the surroundings of the refinery needs to be taken into account, since the number of persons present influences the societal risk. Fig. 12.2 presents the population distribution considered for the test case.

For the present QRA no distinction was made between day and night population densities. Rather, it was assumed that the population distribution does not

Legend	Population density range (km^{-2})	Representative population density (km^{-2})
	0–250	125
	250–1000	625
	1,000–5,000	3,000
	5,000–10,000	7,500
	10,000–113,318	62,000

FIGURE 12.2 Population Distribution per km^2 Used for the Case Study

vary between day and night. Two weather classes were considered: (1) D5: neutral atmosphere with 5 m/s wind, which occurs during both day- and night-time, and (2) F1.5: very stable atmosphere with 1.5 m/s, which occurs only during the night. The probability of a specific wind direction is random, that is, the wind rose is circular. The standard accident scenarios and related frequencies adopted for the QRA of the given test case are reported in Table 12.1.

Normally, two pipe-related scenarios are considered for QRA's: a full-bore rupture and a small leak. Depending on the configuration of the pipe, a full-bore rupture could result in a two-sided outflow, that is, outflow determined by the upstream equipment/conditions and back flow from the downstream equipment. In order to consider pipe failure correctly, detailed information on its layout, operating conditions, and implemented safety measures is necessary. Unfortunately, information at this level of detail is not available and therefore pipe-related accident scenarios as defined in the *Purple Book* were not taken into account. However, since pipes and pipelines

Table 12.1 Standard Scenarios and Associated Frequencies for Industrial Accidents

	Frequency (Year^{-1})	
Scenario	Atmospheric Vessels (Single Containment)	Pressurized Vessels
Instantaneous release of the complete inventory	5×10^{-6}	5×10^{-7}
Continuous release of the complete inventory in 10 min at a constant rate of release	5×10^{-6}	5×10^{-7}
Continuous release from a hole with an effective diameter of 10 mm	10^{-4}	10^{-5}

Table 12.2 Standard Scenario and Frequencies for Above-Ground Pipes With Diameter D

	Frequency (Year^{-1})		
Scenario	D < 75 mm	75 mm < D < 150 mm	D > 150 mm
Full-bore rupture	10^{-6}	3×10^{-7}	10^{-7}

might be more vulnerable to natural hazards than storage vessels, completely ignoring pipes would result in an underestimation of the impact of natural hazards on the overall risk. For this reason, an additional scenario was defined: the full-bore rupture of a pipe connected to a storage vessel, resulting in the loss of containment of the storage vessel. Backflow from downstream equipment is not considered. The failure frequency of the pipe connection is based on the *Purple Book* and 1-m pipe length (see Table 12.2).

Most tanks are located in large catch basins that encompass several tanks while each tank is located in a dedicated subbasin. This means that small spills remain contained in the subbasin where the leakage occurs without spreading to the other tanks, and for larger leakages the liquid can spread to the neighboring subbasin, but remains contained within the main basin. For the large-leak scenarios (instantaneous and 10-min release), we assumed that the liquid will spread beyond the subbasin. These scenarios have been located at the center of the main basin. For the instantaneous-release scenario we assumed that due to overtopping of the basin wall a surface of 1.5 times the main-basin area is covered by liquid. These assumptions are summarized in Table 12.3. Failure of the bund (basin) walls was not considered. Domino effects were also not taken into account although large pool sizes may cause domino effects upon ignition (e.g., BLEVEs of the LPG spheres).

In case of the impact of a natural event, its hazard and the associated vulnerability of the industrial equipment need to be evaluated. For the specific location of the refinery, earthquake and tsunami hazards were assessed within the STREST project

Table 12.3 Pool Surface Areas

Scenario	Pool Surface Area Considered
Instantaneous release tank	1.5 × main basin
10-min release tank	Main basin
10-mm leak tank	Subbasin

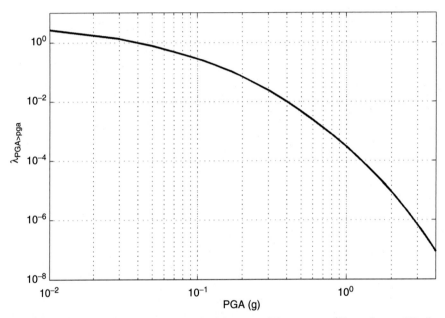

FIGURE 12.3 Hazard Curve in Terms of Mean Annual Frequency of Exceedance of Peak Ground Acceleration (PGA)

(STREST, 2016). The results were expressed in terms of PGA and hw^2 occurrence, where PGA is the peak ground acceleration registered for the earthquake and hw^2 is the product of the height and the velocity of the tsunami wave impacting the equipment. Here, the annual mean frequency of PGA exceedance is shown, for the sake of brevity (Fig. 12.3).

By combining natural-hazard curves and equipment vulnerabilities, new accident scenarios (expressed in terms of loss of containment like for conventional industrial accident analysis) were defined. The final results are listed in Table 12.4, where the representative scenarios and the corresponding annual frequencies are shown. More details on the procedure for Natech QRA can be found in Chapter 7, in the open literature (Salzano et al., 2003; Fabbrocino et al., 2005; Campedel et al., 2008; Lanzano et al., 2013, 2014, 2015; Basco and Salzano, 2016), and in STREST project deliverables (STREST, 2016).

Table 12.4 Accident Scenarios and Frequencies for Stationary Vessels due to Natural-Hazard Impacts (Salzano et al., 2003; Fabbrocino et al., 2005; Campedel et al., 2008; Lanzano et al., 2013, 2014, 2015)

Scenario	Frequency (Year^{-1})		
Earthquake	**Atmospheric Vessels (Single Containment)**	**Pressurized Vessels**	**Pipes**
Instantaneous release of the complete inventory	3.70×10^{-3}	1.16×10^{-9}	—
Continuous release of the complete inventory in 10 min at a constant rate of release	3.70×10^{-3}	1.16×10^{-9}	—
Continuous release from a hole with an effective diameter of 10 mm	7.33×10^{-2}	0	—
Full-bore rupture	—	—	5.56×10^{-2}
Tsunami	**Atmospheric Vessels (Single Containment)**	**Pressurized Vessels**	**Pipes**
Instantaneous release of the complete inventory	1.85×10^{-5} to 3.47×10^{-4}	0	—
Continuous release of the complete inventory in 10 min at a constant rate of release	1.85×10^{-5} to 3.47×10^{-4}	0	—
Continuous release from a hole with an effective diameter of 10 mm	0	0	—
Full-bore rupture	—	—	0
Earthquake + Tsunami	**Atmospheric Vessels (Single Containment)**	**Pressurized Vessels**	**Pipes**
Instantaneous release of the complete inventory	3.7×10^{-3} to 4.05×10^{-3}	1.16×10^{-9}	—
Continuous release of the complete inventory in 10 min at a constant rate of release	3.7×10^{-3} to 4.05×10^{-3}	1.16×10^{-9}	—
Continuous release from a hole with an effective diameter of 10 mm	7.33×10^{-2}	0	—
Full-bore rupture	—	—	5.56×10^{-2}

For earthquake-induced Natech accidents, the release frequencies are location-independent: all atmospheric vessels have the same release frequency, as do the pressurized vessels and pipes. Small releases are not considered for pressurized vessels. The frequencies refer to the likelihood of failure of one or more tanks as a result of the earthquake or tsunami. It is assumed that one catch basin is large enough to hold the volume of the failing tanks and that the catch basin remains intact during and after an earthquake or tsunami. Hence, the pool size is not affected by the

Table 12.5 Representative Substances

Product	Representative Substance
Atmospheric residue	NA
Crude oil	Pentane
Fuel oil	Nonane
Gasoil	Nonane
Gasoline	Pentane
HVGO	NA
Jet fuel/kerosine	Nonane
LPG	Propane
Naphtha	Pentane
Others	Pentane
VAC residue	NA

number of tanks collapsing. Similar to conventional industrial risks, for the earthquake-triggered instantaneous-release scenario we assumed that due to overtopping of the basin wall a surface of 1.5 times the main-basin area is covered by liquid. The consequence analysis was based on the released volume of one tank.

For tsunami-induced releases the location of the vessel determines the release frequency. Only vessels within approximately 250 m from the shoreline will suffer damage due to the tsunami and release their contents. The release frequency is not the same for all vessels. For instance, all pressurized vessels are located further inland away from the shore and will not be affected by the tsunami. No pipe ruptures or small leakages are caused by the tsunami, no matter where the vessels or pipes are located.

Not all hazardous substances present on site were considered individually, and representative substances were used instead. Table 12.5 shows which representative substance was used for each product. Atmospheric residue, heavy vacuum gas oil, and vacuum residue were not considered in the QRA.

12.4 RESULTS AND DISCUSSION

The risk of each separate event was first compared, that is, industrial risk from conventional release sources not related to natural hazards, earthquake- and tsunami-induced risks. Fig. 12.4 presents the LR contours for each of these risks. The black crosses indicate the location of the scenarios.

Conventional industrial risks result in large contours, especially for the lower risk levels (10^{-7} and 10^{-8} year^{-1}). These contours are dominated by the risks related to the LPG storage vessels.

When considering only earthquake-induced risks, the 10^{-7} and 10^{-8} year^{-1} risk contours are smaller. This is due to the lower release frequency for the LPG tanks compared to conventional accident causes. The higher risk levels ($> 10^{-6}$ year^{-1})

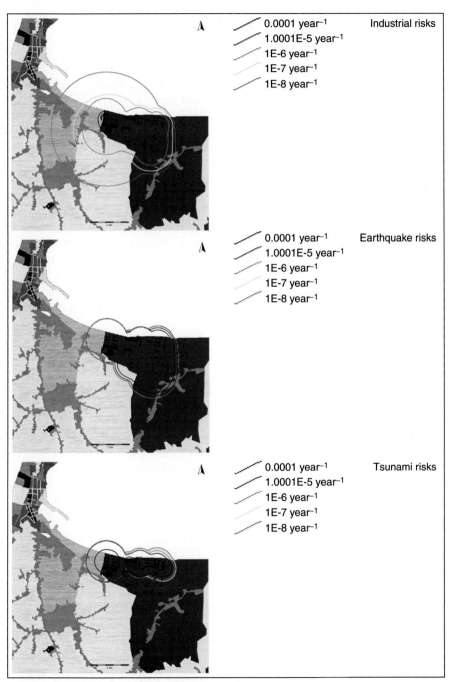

FIGURE 12.4 Locational Risk—Single Causes Only

are dominated by the atmospheric tanks which for earthquakes have a higher release frequency compared to that of conventional industrial risks, resulting in larger 10^{-5} and 10^{-4} year^{-1} contours. The industrial risks on the right side of the site are mainly caused by the atmospheric tanks. When comparing pure industrial risks with earthquake-induced risks, one can observe that in case of an earthquake the risks on the right side of the site have increased by a factor of approximately 1000 and that the 10^{-4} year^{-1} contour is located at almost the same location as the 10^{-7} year^{-1} contour for conventional industrial risks. This is due to the failure frequency of atmospheric tanks being a factor 1000 higher for earthquakes than for failure due to conventional accident causes.

The risks associated with tsunami-induced releases are the smallest of the three release causes. Only atmospheric vessels close to the shore were predicted to result in hazardous-materials releases. Vessels located further away do not pose risks.

Fig. 12.5 shows the LR when considering (1) industrial + earthquake-induced risks; (2) industrial + tsunami-induced risks; and (3) industrial + earthquake- and tsunami-induced risks. The most dominant risks are the industrial and earthquake-induced risks. The figure indicates that for this pilot case, low risks ($< 10^{-6}$ year^{-1}) are dominated by the industrial risks as they are caused by failure of the LPG vessels from causes not related to natural hazards. Earthquake and tsunami do not damage these vessels.

Natural hazards cause an increase in the total risk levels. As the tsunami only damages a limited number of vessels along the shore line, the risk increase is, however, limited. Similar results for earthquakes only and involving a simplified analysis of the earthquake hazard, can be found in Salzano et al. (2003, 2009), Campedel et al. (2008), or Fabbrocino et al. (2005).

Fig. 12.6 shows the societal risk per accident cause. Up to approximately 200 fatalities, Natech accidents from earthquake and tsunami provide a more important contribution due to the higher failure frequency of the atmospheric tanks under these conditions. Larger numbers of fatalities are only caused by industrial risks or earthquakes. This is due to consequences from the failure of the LPG vessels which are not affected by the tsunami due to their location further away from the shore.

12.5 CONCLUSIONS

Natural hazards can play an important role in the total risk of installations with hazardous substances. In this study, the effect of an increased frequency (caused by earthquakes or tsunamis) of a number of hazmat release scenarios on locational and societal risk was assessed.

The impact of natural hazards depends on many (location-specific) factors. For the site analyzed in this study, the tsunami damaged only a limited number of atmospheric storage vessels along the shore line. Hence, the increase of the total risk due to tsunami is limited. Nonetheless, the potential overloading of emergency response should be considered, at least for the tanks along the coastline.

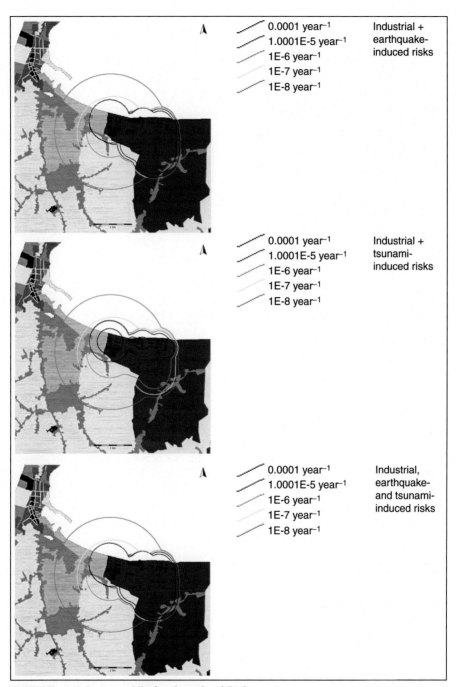

FIGURE 12.5 Locational Risk—Cumulated Risks

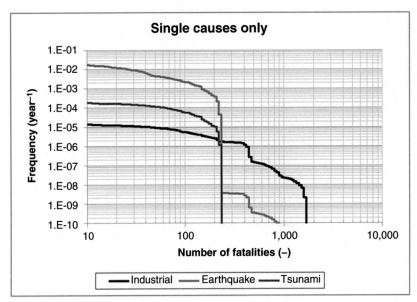

FIGURE 12.6 Societal Risk—Industrial, Earthquake- or Tsunami-Induced

Of more importance is the effect of an earthquake, which significantly increases the failure frequency of atmospheric storage tanks. However, neither an earthquake nor a tsunami considerably increases the failure frequency of and hence the risk posed by pressurized vessels (like LPG spheres). Since for the considered site the risk is largely dominated by the LPG tanks (which fail due to accident causes not related to natural hazards), the impact of these natural events is limited.

This pilot case was performed to show the impact of natural hazards on the outcome of a quantitative risk analysis of an industrial site where hazardous substances are present. The aim was not to perform a detailed QRA of the pilot site (for such an exercise much more detailed information would have been required) but merely to show how (the most common) accident scenarios are affected by an increased release frequency caused by earthquakes and tsunamis.

Other scenarios that can be relevant in case of earthquakes or tsunamis were not evaluated. For instance, failure of multiple tanks was not taken into account. However, multiple and simultaneous releases may result in release volumes that can exceed the catch basins' capacity, and hence lead to larger pool sizes, especially if the catch basins fail. Domino effects were also not considered although cascading events can be more frequent during Natech accidents (cf. Chapter 3 and Section 7.3.2.2). For instance, if a pool of flammable material extends to an area with LPG spheres, BLEVEs may occur. The impact of debris (or large objects like ships) swept onshore with a tsunami is also a source of danger to hazardous industry that was not taken into account in this study. Such phenomena will result in larger impact areas, and hence may increase the number of casualties.

It is interesting to note that with earthquakes and tsunamis having large impact areas, many of the fatalities calculated in the QRA might have occurred also had such an installation not been present, for example, due to building collapse. In other words, the additional number of casualties caused by the damaged industrial installations may very well be much smaller under these conditions. Nevertheless, natural hazards can be an important contributor to the overall risk level of a hazardous installation and should therefore be considered in a facility's risk assessment to ensure appropriate prevention and preparedness.

References

Basco, A., Salzano, E., 2016. The vulnerability of industrial equipment to tsunami, accepted for publication in J. Loss Prev. Process Ind.

Campedel, M., Cozzani, V., Garcia-Agreda, A., Salzano, E., 2008. Extending the quantitative assessment of industrial risks to earthquake effects. Risk Anal. 28, 1231.

Fabbrocino, G., Iervolino, I., Orlando, F., Salzano, E., 2005. Quantitative risk analysis of oil storage facilities in seismic areas. J. Hazard. Mater. 12, 361.

Krausmann, E., Cozzani, V., Salzano, E., Renni, E., 2011. Industrial accidents triggered by natural hazards: an emerging risk issue. Nat. Hazards Earth Syst. Sci. 11, 921.

Lanzano, G., Salzano, E., Santucci de Magistris, F., Fabbrocino, G., 2013. Seismic vulnerability of natural gas pipelines. Reliab. Eng. Syst. Saf. 117, 73.

Lanzano, G., Salzano, E., Santucci de Magistris, F., Fabbrocino, G., 2014. Seismic vulnerability of gas and liquid buried pipelines. J. Loss Prev. Process Ind. 28, 72.

Lanzano, G., Santucci de Magistris, F., Fabbrocino, G., Salzano, E., 2015. Seismic damage to pipelines in the framework of Na-Tech risk assessment. J. Loss Prev. Process Ind. 33, 159.

Ministry of Infrastructure and Environment, 2010. Besluit Externe Veiligheid Buisleidingen (Decree External Safety Pipelines), Dutch Ministry of Infrastructure and Environment.

Salzano, E., Basco, A., Busini, V., Cozzani, V., Renni, E., Rota, R., 2013. Public awareness promoting new or emerging risk: industrial accidents triggered by natural hazards. J. Risk Res. 16, 469.

Salzano, E., Garcia-Agreda, A., Di Carluccio, A., Fabbrocino, G., 2009. Risk assessment and early warning systems for industrial facilities in seismic zones. Reliab. Eng. Syst. Saf. 94, 1577.

Salzano, E., Iervolino, I., Fabbrocino, G., 2003. Seismic risk of atmospheric storage tanks in the framework of Quantitative Risk Analysis. J. Loss Prev. Process Ind. 16, 403.

STREST, 2016. Harmonised approach to stress tests for critical infrastructures against natural hazards, EU Seventh Framework Program, Grant Agreement: 603389. www.strest-eu.org

TNO, 2015. RISKCURVES, Risk assessment software, version 9.0.26, The Netherlands Organization for Applied Scientific Research (TNO). www.tno.nl/riskcurves

VROM, 2005a. Methods for the determination of possible damage to people and objects resulting from releases of hazardous materials. Green Book, PGS 1, Ministry of Housing, Spatial Planning and the Environment, The Hague.

VROM, 2005b. Guidelines for quantitative risk assessment. Purple Book, PGS 3, Ministry of Housing, Spatial Planning and the Environment, The Hague.

VROM, 2005c. Methods for determining and processing probabilities. Red Book, PGS 4, Ministry of Housing, Spatial Planning and the Environment, The Hague.

VROM, 2005d. Methods for the calculation of physical effects. Yellow Book, PGS 2, Ministry of Housing, Spatial Planning and the Environment, The Hague.

Reducing Natech Risk: Structural Measures

A.M. Cruz*, E. Krausmann**, N. Kato†, S. Girgin**
*Disaster Prevention Research Institute, Kyoto University, Kyoto, Japan; **European Commission, Joint Research Centre, Ispra, Italy; †Department of Naval Architecture and Ocean Engineering, Osaka University, Osaka, Japan

Structural prevention and mitigation measures can help prevent damage and hazardous-materials releases at industrial facilities, and contribute to reducing their consequences if releases do occur. This chapter introduces a selection of available structural protection measures for different types of natural hazards.

13.1 INTRODUCTION

Releases of hazardous materials from damaged process or storage equipment pose a substantial threat to human health and the environment. Often, industrial facilities that handle hazardous materials are located in urbanized areas subject to natural-hazard events. In these cases, a Natech event can endanger not only plant personnel, but also residents of the neighboring community. The danger of releases in highly urbanized areas has been explicitly recognized by many developed and developing countries (OECD, 2012).

Limiting industrial development in areas prone to natural hazards is the most efficient way to minimize the danger associated with the potential natural-hazard impact. However, land-use-planning restrictions for existing installations are often very costly and difficult to implement. In this case, supplementary measures are required to protect hazardous facilities. There are various ways in which the risk of Natech accidents can be reduced, including through structural risk-reduction measures, that is, using engineering solutions, as well as through organizational measures. Furthermore, these can be divided into:

- prevention measures—actions or measures that are put in place to reduce the likelihood of damage and occurrence of a hazardous-materials release and
- mitigation measures—actions or measures that are put in place to reduce the impact of hazardous-materials releases if they do occur.

In the following sections we discuss passive and dynamic structural accident prevention and mitigation measures for different types of industrial equipment and building structures, and for different types of natural-hazard events.

Natech Risk Assessment and Management. http://dx.doi.org/10.1016/B978-0-12-803807-9.00013-9

13.2 PREVENTION MEASURES

13.2.1 Earthquakes

In areas of high seismic risk, large earthquakes (M_w 7.0 or greater) pose one of the greatest threats to industrial plants and other infrastructures housing hazardous materials. The examples described in previous chapters highlight that concrete buildings, steel storage tanks, open steel structures, and other equipment present at industrial facilities are vulnerable to earthquake loads. Steel storage tanks are generally classified as anchored or unanchored depending on the restraint provided to the ground. Unanchored storage tanks may be subjected to uplifting and/or sliding motion with the subsequent tearing of connected pipes in case of strong ground motions (Salzano et al., 2003). Welds in steel tanks are sensitive to corrosion and can lead to wide cracks during earthquake events, particularly in the shell/roof and shell/base-plate joint zones. Liquid sloshing in full (or nearly full) steel storage tanks can result in large axial compressive stresses of the tank shell, causing elephant-foot buckling due to the seismic overturning forces. Sloshing of the liquid near the free surface can damage the roof and upper shell of the tanks (Ballantyne and Crouse, 1997; Salzano et al., 2003; Hosseinzadeh and Valaee, 2006).

Damage or collapse of buildings at adjacent structures at process plants can also cause hazardous-materials releases, major process upsets, or even human casualties (Fig. 13.1). Moreover, industrial installations are a combination of buildings and warehouses, industrial equipment, and lifeline systems, such as plumbing, electrical, heating, ventilation, and air-conditioning systems. An earthquake can damage one or a combination of these infrastructures, thereby causing failures in other parts of the facility through these interconnected systems (Cruz, 2014). In fact, often chemical accidents occur not necessarily because of the earthquake itself but due to the secondary effects of the earthquake, such as power outages, loss of water supply (e.g., collapse of water cooling towers, damage to water pipes), lack of process air, damage of critical equipment (e.g., inoperability of boilers), or failure of standard prevention and mitigation measures.

With earthquakes remaining unpredictable despite advances in the natural and engineering sciences, the adoption of appropriate seismic building codes for new plant structures and the retrofitting of older facilities to comply with the latest design codes can help minimize loss of life and property. Building structures related to the operation and administration of a plant or as part of the plant infrastructure (e.g., control rooms, communications, pumping stations, water treatment), or for materials and equipment storage may require special design to ensure that they remain operational following a major disaster.

The impacts of large earthquakes on industrial installations may be severe as the examples of the Kocaeli (cf. Section 2.2) and Wenchuan earthquakes (Krausmann et al., 2010) demonstrated. Lessons from past earthquakes suggest that industrial facilities that adopted earthquake design codes or implemented retrofitting generally performed better (even in earthquakes exceeding design loads) than facilities that had not embraced seismic design. The example of the

FIGURE 13.1 Pipes Collapsed Onto a Building at a Fertilizer Plant During the 2008 Wenchuan Earthquake in China

Photo credit: E. Krausmann.

Great East Japan earthquake in Japan in 2011 (cf. Section 2.3), demonstrated the effectiveness of earthquake prevention measures in Japan through performance-based earthquake design as relatively little damage of major severity occurred due to the earthquake.

Different countries use different seismic building codes. For example, the United States leaves the decision of which building code to adopt to each State, although seismic requirements for new and existing federal buildings exist. The State of California adopted the 2016 California Building Standards Code, which is based on the International Building Code (IBC, 2015). As was mentioned in Chapter 4, chemical accident prevention is regulated in California by the CalARP program. CalARP updated its Guidance for Seismic Assessment in Dec. 2013 (CalARP, 2013), which specifically requires the assessment of:

1. Regulated processes as defined by CalARP Program regulations.
2. Adjacent facilities whose structural failure or excessive displacement could result in the significant release of regulated substances.
3. Onsite utility systems and emergency systems which would be required to operate following an earthquake for emergency reaction or to maintain the

facility in a safe condition (e.g., emergency power, leak detectors, pressure relief valves, etc.).

Most importantly, the seismic assessment guidelines provide several performance criteria that apply to individual equipment items, structures, and systems (e.g., power, water, etc.). These criteria may include one or more of the following:

- maintain structural integrity,
- maintain position,
- maintain containment of material, and
- function immediately following an earthquake.

Japan follows the Building Standard (BS) Law revised in 1981 and the Act for Promoting Earthquake Proof Retrofitting of Buildings enacted in 1995 following the Kobe earthquake (Cruz and Okada, 2008). It is generally agreed that engineered buildings in Japan are designed for greater strength and stiffness than similar buildings in the United States and tend to be more resilient (Whittaker et al., 1998). One important aspect of the revised BS law is that it uses performance-based standards where the building is required to satisfy performance criteria with respect to materials, equipment, and structural methods (Japan External Trade Organization, 2005). As explained in Chapter 4, the seismic code was modified to account for lessons learned from the Great East Japan Earthquake in 2011. Furthermore, in Mar. 2016 the Institute for Disaster Mitigation of Industrial Complexes at Waseda University in Tokyo published Guidelines for Earthquake Risk Management at Industrial Parks (IDMIC, 2016). The Guidelines are similar to the CalARP seismic guidance document emphasizing various performance levels, addressing soil problems and structural design issues, as well as prevention and mitigation measures. It differs from the CalARP in that the Japanese guidelines are meant for area-wide assessment at industrial agglomerated areas.

13.2.1.1 *Storage Tanks*

Storage tanks may be directly damaged by both short- and long-period earthquake ground motions. Typical failure modes of storage tanks during past earthquakes include shell buckling, roof damage, anchorage failure, tank-support system failure, foundation failure, hydrodynamic pressure failure, connected pipe failure, and manhole failure (ALA, 2001; Salzano et al., 2003). Storage tanks may suffer damage due to earthquake-induced liquid sloshing and soil liquefaction. Liquid sloshing and the resulting dynamic loading on the tank wall need to be taken into account in the design of storage tanks in earthquake-prone areas. During the Great East Japan earthquake most storage tanks performed well except along the northwest Japan Sea coast and the Tokyo Bay area, which experienced long-period strong ground motions. This caused liquid sloshing of oil in storage tanks with resultant sinking of floating roofs and other damage, such as failure of pontoons (Zama et al., 2012).

The earthquake prevention measures adopted should consider the tank's intended use, size, structure type, materials, design lifetime, location, and environment in order to assure life safety and to maintain their essential functions following an earthquake. The Architectural Institute of Japan, which published the 10th edition of "Design Recommendation for Storage Tanks and Their Supports With Emphasis on Seismic Design" (AIJ, 2010), noted that the trend in recent years has been for larger storage tanks, and as such the seismic design for these larger tanks has become more important to guarantee public safety and environmental protection. In addition to the seismic design of the tank structure and foundation, the appropriate Japanese codes (e.g., Fire Safety Code) and standards (e.g., API Standard 650 for welded steel oil storage tanks) should be followed.

Anchoring of above-ground storage tanks can prevent horizontal sliding, tank uplifting, and tank overturning. Anchor and anchor-chair design must be considered carefully to avoid damage between anchor chair and tank shell due to excessive rigidity of anchors (Sakai et al., 1990). Furthermore, some industry guidelines, such as the Process Industry Practice (PIP, 2005) Guidelines for Tank Foundation Designs, do not recommend the use of anchors on large cylindrical storage tanks. Bakhshi and Hassanikhah (2008) studied the performance of anchored and unanchored storage tanks during the Kobe earthquake in 1995 and found that the main differences are related to uplifting phenomena. The authors found that tall and medium storage tanks are more sensitive to anchorage conditions than larger, broader storage tanks.

Damage to tank flanges and connected pipes can be reduced by the use of flexible pipe coupling and flexible pipes as shown in Fig. 13.2.

Damage of support legs of spherical storage tanks during the Great East Japan earthquake demonstrated the need for improved design and strengthening of leg braces. Furthermore, the need to consider situations where the equipment may be under higher stresses than those experienced during normal operation, e.g., during equipment maintenance and checks, need to be reviewed, and factored into the design of these storage tanks if needed.

13.2.1.2 Pipework and Pipelines

With respect to overland transport pipelines, the cheapest and most effective way to protect the pipeline and its network components (pump/metering stations, valve sites, terminal/tank farms) is adequate siting to keep the vulnerable equipment away from earthquake-prone areas. As this may not always be possible, the risk of accidents in such situations can be significantly reduced by the careful selection of pipeline routes, the line pipe's orientation with respect to fault lines, and hazard-aware choices for the siting of critical components of the pipeline network (Yokel and Mathey, 1992).

In addition to sensible siting, design safety is the most important pipeline protection mechanism. It relies on the availability and implementation of modern design standards, and it includes the use of resistant pipe materials and novel techniques for the strengthening of joints to better resist seismic loading. In areas of permanent ground deformation induced by liquefaction or at fault crossings, additional measures are required to effectively protect a pipeline. Adjusting the orientation of the line pipe with respect to the fault direction or using low-density backfill material at the trench are common practices in such situations (Girgin and Krausmann, 2015; STREST, 2014).

FIGURE 13.2

Flexible Pipe Coupling (A) and Flexible Steel Pipe (B) on Large Oil Tanks (20 × 40 m)

Photo credit: A.M. Cruz.

The Trans-Alaska oil pipeline is an excellent example of how engineering solutions can successfully protect pipelines from even severe seismic activity. The Trans-Alaska pipeline dates back to the 1970s when it was built according to stringent earthquake design specifications to accommodate the possibility of a M = 8 earthquake from the Denali Fault which the pipeline crosses. The implemented earthquake design was put to the test in Nov. 2002 when the fault ruptured during a M = 7.9 earthquake. The strong shaking forces damaged a number of the pipeline's

FIGURE 13.3 Engineered Seismic Protection Measures Implemented for the Trans-Alaska Pipeline at the Denali Fault Crossing

Adapted from USGS (2003), Courtesy of the US Geological Survey.

supports near the fault area, but the pipeline easily accommodated the 4.3-m horizontal and 0.8-m vertical shift at the fault crossing without breaking (USGS, 2003). Having been aware of the possibility of strong earthquakes in the area, pipeline designers had given the line pipe Teflon shoes with which it could slide on long horizontal beams, thereby allowing the pipe to move and giving it flexibility under seismic stress (Fig. 13.3). The overall cost of implementing this measure [about 3 million US$ (in 1970 US$)] was considered significantly below the potential economic losses due to lost revenue and repair costs, as well as environmental cleanup, had the pipeline ruptured.

In case an earthquake occurs, quick operator action might be necessary, such as reducing the flow in the pipeline or shutting it down completely to reduce the stresses on the pipeline wall. In this context, Griesser et al. (2004) discuss the installation of strong-motion detectors on pipelines in seismic areas. Based on the information provided by these detectors, control signals can be issued to support quick shutdown or other types of preventive action.

13.2.2 Tsunami and Coastal Storm Surge

Buildings, storage tanks, and pipelines located in coastal areas subject to tsunami hazards may be vulnerable to wave impact and flooding. Site conditions in the run-up zone will determine the depth of tsunami inundation, water flow velocities, the presence of breaking wave or bore conditions, debris load, and warning time, and can vary greatly from site to site (NTHMP, 2001). The vulnerability of buildings to tsunami loads will depend on several factors including number of floors, the presence of open ground floors with movable objects, building materials, age and design, and building surroundings, such as the presence of barriers (Dominey-Howes and Papathoma, 2007).

The National Tsunami Hazard Mitigation Program (NTHMP, 2001, 2013) in the United States recommends four basic techniques that can be applied to buildings and other infrastructure to reduce tsunami risk, including:

- Avoiding development in inundation areas: This is the most effective prevention strategy but not always possible particularly for existing buildings.
- Slowing techniques: These include the use of specially designed forests, ditches, slopes, and berms which can slow and drain debris from waves.
- Steering techniques: These are used to guide tsunamis away from vulnerable structures and people by placing structures, walls, and ditches and using paved surfaces that create a low-friction path for water to follow.
- Blocking water forces: This technique consists in building hardened structures, such as break walls and other rigid construction that can block the force of the waves.

Until recently, there were no tsunami-specific building codes. Structural design to protect buildings in tsunami-prone regions is generally based on loading due to riverine floods and storm waves, providing little guidance for loads specifically induced by tsunami effects on coastal structures (Yeh et al., 2005). Recently, a new chapter on "Tsunami Loads and Effects" for the 2016 edition of the American Society of Civil Engineer's ASCE 7-16 Standard "Minimum design loads and associated criteria for buildings and other structures" was introduced in the United States. According to Chock (2015), the ASCE 7-16 Tsunami Loads and Effects chapter will become the first of its kind in the United States for use in the states of Alaska, Washington, Oregon, California, and Hawaii. Chock (2015) explains that the new ASCE 7 provisions implement a unified set of analysis and design methodologies that are consistent with probabilistic hazard analysis, tsunami physics, and structural target reliability analysis. The approach developed results in the first unification of tsunami hazard mapping for design and reflects a modern approach of performance-based engineering.

In Europe there has been an effort at the European Community level, through the Tsunami Risk and Strategies for the European Region (TRANSFER) project, cofunded under the European Union 6th Framework Programme, to improve the knowledge of tsunami processes and risks in tsunami-prone regions particularly in areas, such as Southern Italy, Southern Spain, and Greece. Efforts have centered on modeling hazards, hazard mapping, and vulnerability assessment of critical and essential infrastructure systems [see, e.g., Cruz et al. (2009) for a study on tsunami impact at a refinery in the south of Italy]. One of the key goals of the project was the development of strategies and policies to manage, mitigate, and deal with risks stemming from future tsunami hazards.

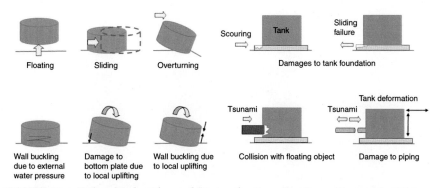

FIGURE 13.4 Failure Modes Observed During the Great East Japan Tsunami in 2011

Adapted from Ibata et al. (2013).

13.2.2.1 Storage Tanks

Storage tanks are vulnerable to tsunami loads. Ibata et al. (2013) report that damage to many cylindrical oil storage tanks during the Great East Japan earthquake and tsunami occurred due to sliding, floating, overturning, steel wall buckling, and collapse due to the forces of the tsunami. Fig. 13.4 gives an overview of the various failure modes of tanks under tsunami loading, while Fig. 13.5 presents photos of two tanks damaged by the Great East Japan tsunami. Of the 167 tanks reported damaged, 120 tanks had capacities of less than 500 kL. There was no damage to tanks with capacities larger than 10,000 kL, but the pipework for 27 of these tanks was affected. Slowing, steering, and blocking water techniques may be used for storage-tank protection. Their structural design and heights should be carefully evaluated based on a tsunami-hazard assessment. During the Great East Japan tsunami, earthen dikes around storage tanks at a refinery in Sendai were overtopped. Although the tanks did not float off their foundation, debris impacts caused damage to connected pipes resulting in oil releases. Since the earthquake and tsunami, the earthen dikes have been reconstructed and their height increased. Fig. 13.6 shows the reconstruction of a damaged earthen dike at the refinery in Sendai, Japan.

Submersion and subsequent buoyancy are of particular concern for storage tanks that may float off their foundations, thereby tearing pipe connections and resulting in the release of possibly flammable and/or toxic materials. According to Ibata et al. (2013) damage to storage tanks in the Great East Japan tsunami increased with inundation depth higher than 3 m. The authors reported that for inundation depths between 3 and 5 m damage to attached piping and tank body were documented. At inundation depths of over 5 m, most of the storage tanks were damaged.

Research is ongoing to improve the protection of storage tanks during tsunami. Examples include the development of a method to reduce damage by tsunami through design and the application of flexible pipes to reduce tsunami flow loads (Okubayashi et al., 2016; Tar et al., 2016), and tsunami impact minimization based on storage tank distribution.

FIGURE 13.5 Examples of Damage to Oil Storage Tanks in Miyagi Prefecture During the Great East Japan Tsunami in 2011

Photo credit: C. Scawthorn.

13.2.2.2 Pipework and Pipelines

Pipework and pipelines are vulnerable to direct tsunami impact and debris loads, as well as hydrostatic and buoyancy loads which may result in pipes floating off and breaking. Overland pipelines were especially vulnerable to tsunami debris impact during the Great East Japan tsunami resulting in several oil spills (Ibata et al., 2013). Commonly used guidelines suggest that a pipe should not suffer any displacement from wave action with a 5-year return period but can experience minor displacement from wave action with a 50-year return period. Major displacement is possible for wave loading with a 100-year return period, but nevertheless the pipe should

FIGURE 13.6 Reconstruction of an Earthen Dike at a Refinery in Japan After the 2011 Tohoku Earthquake

Photo credit: A.M. Cruz.

never collapse. Submerged pipes or pipes near the coastline in tsunami-prone areas should be anchored or braced, and have ballast weights to avoid flotation. Support structures should take into account tsunami wave scouring and soil erosion, or soil liquefaction. Furthermore, lessons from the Great East Japan tsunami showed the need to protect pipelines from debris impact.

13.2.2.3 *Other*

In addition to implementing equipment-specific structural protection measures, the tsunami risk to hazardous installations in coastal areas can be reduced by lowering the impact forces of tsunami waves. This can be achieved by putting into place offshore breakwalls or other types of barriers onshore (Ergin and Balas, 2006; Jayappa, 2008; Maheshwari et al., 2005). These physical barriers could also keep tsunami-driven debris from washing into the plant. In locations where a facility is not protected by external tsunami barriers, it is advisable to take measures to avoid wave-load damage and water intrusion for all structures containing hazardous substances and all systems that are critical for the safety of the installation (Cruz et al., 2011).

13.2.3 Floods

Flood loads are similar to tsunami loads and include hydrostatic, buoyant, hydrodynamic, and breaking-wave forces, as well as impact loading which results from floating debris (Yeh et al., 2005). Buildings located in river basins and near large water bodies may be subject to flood loads. Similar to tsunamis, flood protection measures include avoiding building in flood-prone areas, particularly within the 100-year flood plain, water proofing of buildings, and slowing, steering, and blocking techniques. Elevation of buildings or important building components above the 100-year flood contour level can protect building functionality and contents. The United States Army Corps of Engineers (USACE) has done extensive work in flood mitigation and control. The report "Flood proofing techniques, programs, and references," prepared by the USACE, presents a comprehensive review of flood-proofing techniques (USACE, 1997).

Most wealthier nations (e.g., the United States, Germany, Italy, Spain, France) as well as many developing countries (e.g., Mexico, Colombia) limit or prohibit development in the 100-year flood plain. However, the law generally applies to new construction. Thus, existing buildings located within the 100-year flood plains may not be sufficiently protected. Furthermore, political pressure and corruption sometimes result in concession of building permits or illegal development in flood-prone areas (Sierra, 2005; Santander, 2010).

In Japan, the Ministry of Land, Infrastructure and Transport established the comprehensive Flood Control Measures, which consolidate the combined use of facilities to maintain the water-retaining and retarding functions of river basins, the creation of incentives to use land safely and to build flood-resistant buildings, and the establishment of warning and evacuation systems for both tsunami and riverine flooding. Development in high flood-risk areas (within the 100-year flood plain) is regulated through land-use planning controls and flood plain zoning. In addition, the Building Standard law provides provisions for flood proofing of engineered structures, including construction of seawalls and other barriers to protect port terminal facilities from flooding, storm surge, and tsunami waves (Cruz and Okada, 2008).

13.2.3.1 Storage Tanks

Postaccident analyses showed that storage tanks are particularly vulnerable to flood impact. The main damage and failure mechanisms are buoyancy, water drag, and debris impact (cf. Section 3.5). Adequate anchoring with bolts or other types of restraining systems should effectively prevent tanks and other equipment from floating off their foundations under most flood or storm-surge conditions. Another risk-reduction measure is the filling of empty tanks with water in preparation for a flood situation to avoid tank floating and subsequent displacement. This measure is, however, controversial as product residues might still be in the tank, and safety procedures are required to avoid contamination or reaction of the water with the product. Consequently, the implementation of these procedures requires some lead time and, therefore, reliable early warning. Instead of water, a specified amount of product could be left in the tanks at all times, thereby increasing its weight and decreasing the risk of buoyancy. However, should the tank fail, the consequences could be more severe (Krausmann et al., 2011).

Storage tanks are commonly surrounded by containment dikes or concrete walls which retain accidental releases from the tanks. While these catch basins are not designed to keep the floodwaters out, they can provide some protection provided that they are not overtopped or that erosion from the flooding has not compromised the structural integrity of the dikes.

The risk of debris impacts on vulnerable equipment and the associated hazardous-materials releases can be controlled by creating barriers that steer the floodwaters away from an industrial plant. External barriers, such as earthen berms, sheet pile, or concrete walls, can contribute to keep flood-driven debris from washing into a facility (Cruz and Krausmann, 2013).

13.2.3.2 Pipework and Pipelines

There are various standards or codes to ensure that pipelines are able to withstand anticipated external pressures and loads that will be imposed on the pipelines after installation. Furthermore, most codes will provide guidance on the number of shutoff valves to be installed at intervals ranging from somewhere between 5 and 30 km, depending on the population density or presence of sensitive areas along the route of the pipeline. Pipelines should be protected from floods and flash floods that may result in pipe displacement or cause the pipe to sustain abnormal loads. Pipelines installed in a navigable river, stream, or harbor should be buried and have a minimum cover of soil or consolidated rock (NTSB, 1996). Most importantly, a detailed flood risk assessment should accompany any pipeline design to ensure that the maximum flood-hazard risks have been considered in the design, installation, management, and monitoring of the pipeline.

13.2.3.3 Other

Water intrusion in hazardous or auxiliary equipment can cause short circuits or power loss which could trigger or exacerbate a major accident. Similar to tsunamis, also in the case of floods or storm surge, safety-critical systems in a hazardous installation need to be protected from wave-load damage and water intrusion to guarantee their continued functioning. This can be achieved by waterproofing of vulnerable equipment and systems. The implementation of safe equipment design that makes use of, for example, interlocks, fail-open or fail-closed valves contributes toward ensuring safe emergency shutdown in situations in which onsite power is lost due to flooding (Krausmann et al., 2011).

It has been observed during past Natech accidents that waste oil in a plant's drainage system can be lifted by the floodwaters and be dispersed after stratification on the water surface. Upon contact with an ignition source, which can be a hot plant unit or a lightning strike, major fires and/or explosions can be sparked (Cruz et al., 2001). In areas prone to flooding, including those where a rise in groundwater level is common during periods of long, sustained rainfall, the drainage systems for waste flammable substances and surface run-off water should be segregated.

13.2.4 High Winds

Buildings, storage tanks, and other structures may be subject to wind damage, particularly storm-induced winds, hurricane winds, and tornadoes. Engineering design codes are used to insure that buildings and structures are constructed to withstand particular wind speeds depending on the characteristics of each region. In the United States, the American Society of Civil Engineers (ASCE) provides guidelines for the design and calculation of wind loads in the design standard ASCE 7-05 "Minimum Design Loads for Buildings and Other Structures" (ASCE, 2006). ASCE

7 requires design for the 50-year wind speed with an importance factor for critical infrastructures and industrial facilities containing hazardous materials. This results in the equivalent of a 500-year wind speed for these structures (Steinberg, 2004).

It is important to note that very often wind damage to buildings is due to failure of roofing materials, doors, and windows. These failures, which are often less expensive to prevent or mitigate, lead to weather penetration and damage (Heaney et al., 2000).

In Japan, wind loads are addressed using a performance-based approach. The requirements for building structures in areas subject to high winds in Japan are given by the Wind Load Provisions of the Building Standard law and Building Control System (Cruz and Okada, 2008). These requirements are classified into three categories: life safety, damage prevention, and continuous normal operation. Each of these categories assumes a specific level of load/forces (Hiraishi et al., 1998). Critical facilities and essential building structures will require that they remain operational after being exposed to high wind loads.

13.2.4.1 Storage Tanks

Wind loads on storage tanks include wind pressure on vertical projected areas of cylindrical surfaces and uplift pressures on horizontal projected areas of conical or curved surfaces and roofs. The recently updated design standard ASCE 7-16, "Minimum Design Loads and Associated Criteria for Buildings and Other Structures" provides guidelines for the design and calculation of wind loads on storage tanks (ASCE, 2016).

13.2.5 Lightning

Lightning is a common accident trigger in processing and storage activities (Krausmann and Baranzini, 2012; Rasmussen, 1995). With climate change and the predicted increase in the frequency of severe hydrometeorological hazards, lightning hazards are expected to become more pronounced in the future (IPCC, 2007).

The main purpose of lightning protection is to keep lightning away from flammable and explosive substances, and avoid sparking and flashovers, as well as overheating in conductors. Bouquegneau (2007) emphasizes that in the oil and gas industry, lightning protection systems of class I or even I+ should be adopted in sensitive areas to ensure high safety levels. A number of common protection measures and systems, such as grounding of equipment, lightning rods, or circuit breakers, are available. It has, however, been found that these measures may not prevent equipment damage or failure and the ignition of flammable substances effectively (Renni et al., 2010; Goethals et al., 2008; EPA, 1997). Moreover, lightning protection measures and systems require regular inspections and maintenance as they tend to deteriorate due to chemical corrosion, weather-related effects, and mechanical damage. Protection measures in poor conditions are not effective in preventing an accident.

13.2.5.1 Storage Tanks

Lightning is a frequent source of fires in storage tanks containing flammable substances (Renni et al., 2010; Chang and Lin, 2006). The rim seal of atmospheric floating-roof tanks has been identified as the most likely point of ignition during a lightning strike. Regular checks of the rim seal and maintaining it in good condition will limit the escaping of flammable vapors and hence the risk of ignition during a lightning storm.

The International Standard on protection against lightning (IEC, 2006) indicates that tanks are essentially self-protecting provided they are continuous metallic containers with a minimum shell thickness that depends on the metal the tank is made of (e.g., 4 mm for steel tanks). Additional protection measures might be required for instrumentation and electric systems associated with the operation of these tanks. Bouquegneau (2007) notes that measures for lightning protection should be taken in accordance with the type of tank. Isolated tanks and containers should be earthed at least every 20 m.

The situation is somewhat different for floating-roof tanks containing flammable substances. For more effective lightning protection, the roof should be bonded to the tank shell, and tank seals and shunts that safely conduct stray currents to the ground should be designed with the objective to minimize the risk of ignition. This includes the determination of the optimum number of tank shunts and their location. For floating roof tanks, multiple shunt connections at 1.5-m intervals around the roof perimeter are recommended (Bouquegneau, 2007). Interestingly, in some cases tank shunts have been found to actually increase the risk of fires during lightning storms, as they are a source of sparking when hit by lightning (LEC, 2006).

Currently, no consolidated methodology is available for analyzing the risk of lightning impacts at hazardous installations. However, first attempts in this direction have been made by developing a quantitative methodology for determining the lightning capture frequency of hazardous equipment and the associated damage potential (Necci et al., 2014, 2013). These studies also discussed the benefits of selected types of lightning protection systems and how their positioning onsite can influence risk-reduction efforts.

13.2.5.2 Pipework and Pipelines

The International Standard on protection against lightning also defines protection measures for overland transport pipelines. For instance, it recommends that above-ground metal pipelines should be earthed every 30 m (IEC, 2006). For pipeline stations, surge-protection devices should be implemented to prevent disturbances in control systems. Cathodic protection systems, implemented to reduce the risk of pipe corrosion by establishing a pipe-ground voltage differential, are generally safeguarded against surges and lightning currents. Several incidents suggest, however, that lightning can overcome this system which causes concern as to the effectiveness of corrosion protection.

13.2.5.3 Other

Lightning can cause onsite power blackouts and power dips which can upset processes, affect electrically operated safety systems, and as a consequence lead to loss of containment. It is crucial to identify the dangerous conditions which can result from a power loss to be able to prioritize the processes that should receive emergency power from internal backup systems. It should also be considered to shut down highly hazardous processes under these conditions, blow-down pressurized equipment, and put process units into safe mode. During the starting-up of the installation in the wake of a lightning storm, processes need to be monitored carefully to detect possible malfunctions that could threaten plant safety, as early as possible (Krausmann et al., 2011).

13.3 MITIGATION MEASURES

The design and implementation of mitigation measures to reduce the impact of hazardous-materials releases concurrent with natural disasters requires a careful risk assessment to make certain that these remain functional or operational following a natural disaster. Typical mitigation measures include containment walls/dikes around storage tanks to contain any liquid release and to limit the spill surface area for vaporization in case of volatile substance, the installation of oil-spill detectors and automated emergency shutoff valves to decrease spill quantities, water cannons and foaming systems with in situ and easily accessible (e.g., next to each tank) foam stocks to insulate the spill surface and prevent vaporization, water curtains to wash out toxic gas releases, water sprinkler systems around and over the storage tanks for fighting fires and to cool off tanks in the case of fire in nearby area (Fig. 13.7),

FIGURE 13.7 Retrofitting of a Storage Tank Affected by the 1999 Kocaeli Earthquake in Turkey With a Sophisticated Sprinkler System

Photo credit: S. Girgin.

FIGURE 13.8 Damaged Retention Wall During the 2008 Wenchuan Earthquake

Photo credit: A.M. Cruz.

fire walls to protect control rooms or other sensitive areas (e.g., residential areas), a sufficient number of fire hydrants, above-ground water pipelines with pumps and inlets for external water feed to reduce the risk of pipe breaks due to ground displacement, and high-capacity backup power generators in the case of power outages, etc. (Girgin, 2011; Steinberg and Cruz, 2004). The effectiveness of the existing protection measures should be tested periodically to ensure that they function properly during natural-hazard conditions. Wherever possible, multiple mitigation measures should be implemented to prevent the escalation of the accident.

Unfortunately, natural hazards are still commonly overlooked in the design of engineered protection measures and systems at hazardous facilities. Damage to safety and mitigation measures can render them inoperable and hence unable to perform their functions (Fig. 13.8). Past earthquakes have shown that containment dikes may fail during strong ground motion (Steinberg and Cruz, 2004; Ibata et al., 2013). The use of liners inside containment dikes or walls to prevent leakage in case of tank rupture and dike/wall break will provide added protection. Nevertheless, during the Great East Japan earthquake, soil liquefaction and large ground displacement caused damage to dike walls as well as dike liners (Ibata et al., 2013). Thus, seismic design should also be applied to containment dikes/walls. Moreover, all critical active and passive safety barriers (e.g., water curtains/deluges, containment dikes, retention walls) in a hazardous installation need to be designed to withstand the forces of relevant natural-hazard loads (e.g., from earthquakes, tsunamis, floods, etc.). The use of automatic shutdown systems activated by sensors in the wake of natural-hazard loading may prevent releases if plant units have been damaged during a natural event.

Damage to utilities is a generally occurring and major problem during natural hazards. Although emergency- and backup systems for electricity and water supply are usually available at large plants, they might be inadequate to meet the high demand in case of simultaneous multiple Natech accidents. Backup power generators designed not only to maintain lighting, but sufficiently powerful for the operation of critical equipment and/or other plant operations should be considered and planned for.

References

AIJ, 2010. Design recommendation for storage tanks and their supports with emphasis on seismic design, Sub-Committee for Design of Storage Tanks, 2010 ed. Architectural Institute of Japan. http://www.aij.or.jp/jpn/databox/2011/storagetanks2010edition.pdf

ALA, 2001. Seismic Fragility Formulations for Water Systems, Part 1—Guideline. American Lifeline Alliance, http://www.americanlifelinesalliance.com/pdf/Part_1_Guideline.pdf.

ASCE, 2006. ASCE 7-05 Minimum Design Loads for Buildings and Other Structures. American Society of Civil Engineers, Reston, Virginia.

ASCE, 2016. ASCE 7-16 Minimum Design Loads and Associated Criteria for Buildings and Other Structures. American Society of Civil Engineers, Reston, Virginia.

Bakhshi, A., Hassanikhah, A., 2008. Comparison between seismic responses of anchored and unanchored cylindrical steel tanks. In: Proceedings of the Fourteenth World Conference on Earthquake Engineering, Beijing, China, October 12–17.

Ballantyne, D., Crouse, C., 1997. Reliability and Restoration of Water Supply Systems for Fire Suppression and Drinking Following Earthquakes. National Institute of Standards and Technology (NIST), Gaithersburg, Maryland.

Bouquegneau, C., 2007. Lightning protection of oil and gas industrial plants, In: Proceedings IX International Symposium on Lightning Protection, Foz do Iguaçu, Brazil, November 26–30.

CalARP, 2013. Guidance for California Accidental Release Prevention (CalARP) Program Seismic Assessments. CalARP Program Seismic Guidance Committee, California Governor's Office of Emergency Services, Mather.

Chang, J.I., Lin, C.C., 2006. A study of storage tank accidents. J. Loss Prev. Process Ind. 19, 51.

Chock, G.Y.K., 2015. The ASCE 7 Tsunami loads and effects design standard. In: Proceedings Structures Congress 2015, Portland, Oregon, April 23–25.

Cruz, A.M., 2014. Managing infrastructure, environment and disaster risk. In: López-Carresi, A., Fordham, M., Wisner, B., Kelman, I., Gaillard, J.C. (Eds.), Disaster Management: International Lessons in Risk Reduction, Response and Recovery. Earthscan–Routledge, London.

Cruz, A.M., Franchello, G., Krausmann, E., 2009. Assessment of tsunami risk to an oil refinery in Southern Italy, EUR 23801 EN, European Communities.

Cruz, A.M., Krausmann, E., 2013. Vulnerability of the oil and gas sector to climate change and extreme weather events. Clim. Change 121, 41.

Cruz, A.M., Krausmann, E., Franchello, G., 2011. Analysis of tsunami impact scenarios at an oil refinery. Nat. Hazards 58, 141.

Cruz, A.M., Okada, N., 2008. Consideration of natural hazards in the design and risk management of industrial facilities. Nat. Hazards 44, 227.

Cruz, A.M., Steinberg, L.J., Luna, R., 2001. Identifying hurricane-induced hazardous materials release scenarios in a petroleum refinery. Nat. Hazards Rev. 2 (4), 203.

Dominey-Howes, D., Papathoma, M., 2007. Validating a tsunami vulnerability assessment model (the PTVA Model) using field data from the 2004 Indian Ocean tsunami. Nat. Hazards 40, 113.

EPA, 1997. Lightning Hazard to Facilities Handling Flammable Substances (EPA 550-F-97-002c). United States Environmental Protection Agency, Washington, DC.

Ergin, A., Balas, C.E., 2006. Damage risk assessment of breakwaters under tsunami attack. Nat. Hazards 39, 231.

Girgin, S., 2011. The Natech events during the 17 August 1999 Kocaeli earthquake: aftermath and lessons learned. Nat. Hazards Earth Syst. Sci. 11, 1129.

Girgin, S., Krausmann, E., 2015. Lessons learned from oil pipeline Natech accidents and recommendations for Natech scenario development—Final Report, EUR 26913 EN, European Union.

Goethals, M., Borgonjon, I., Wood, M., 2008. Necessary measures for preventing major accidents at petroleum storage depots: key points and conclusions. Seveso Inspections Series, vol. 1, EUR 22804 EN, European Commission, Ispra; Federal Public Service Employment, Labour and Social Dialogue, Brussels.

Griesser, L., Wieland, M., Walder, R., 2004. Earthquake detection and safety systems for oil pipelines. Pipeline Gas J 38.

Heaney, J., Peterka, J., Wright, L., 2000. Research needs for engineering aspects of natural disasters. J. Infrastruct. Syst. 6 (1), 4.

Hiraishi, H., Teshigawara, M., Fukuyama, H., Saito, T., Gojo, W., Fujitani, H., Okawa, I., Okada, H., 1998. New framework for performance based design of building structures—design flow and social system. In: Proceedings of the Thirtieth Joint Meeting of the US-Japan cooperative program in natural resources, Panel on wind and seismic effects, NIST SP 931.

Hosseinzadeh, N., Valaee, A., 2006. Seismic vulnerability analyses of cylindrical steel above ground tanks in an oil refinery complex. In: Proceedings of the First European Conference on Earthquake Engineering and Seismology (ECEES), Geneva, Switzerland, September 3–8.

Ibata, T., Nakachi, I., Ishida, K., Yokozawa, J., 2013. Damage to storage tanks caused by the 2011 Tohoku earthquake and tsunami and proposal for structural assessment method for cylindrical storage tanks. In: Proceedings of the Seventeenth International Conference & Exhibition on Liquefied Natural Gas (LNG 17), Houston, Texas, April 16–19.

IBC, 2015. International Building Code. International Code Council®, Washington, DC.

IDMIC, 2016. Guidelines for Earthquake Risk Management at Industrial Complexes. Institute for Disaster Mitigation of Industrial Complexes, Waseda University, Tokyo, Japan.

IEC, 2006. Protection against lightning, IEC 62305 International Standard, first ed., International Electrotechnical Commission.

IPCC, 2007. Climate change 2007 impacts, adaptation and vulnerability. Cambridge University Press, Cambridge and New York.

Japan External Trade Organization, 2005. Amended Building Standard Law, Japan.

Jayappa, K.S., 2008. Coastal problems and mitigation measures including the effects of tsunami. Curr. Sci. 94 (1), 14.

Krausmann, E., Baranzini, D., 2012. Natech risk reduction in the European Union. J. Risk Res. 15/8, 1027.

Krausmann, E., Cruz, A.M., Affeltranger, B., 2010. The impact of the 12 May 2008 Wenchuan earthquake on industrial facilities. J. Loss Prev. Process Ind. 23, 242.

Krausmann, E., Renni, E., Campedel, M., Cozzani, V., 2011. Industrial accidents triggered by earthquakes, floods and lightning: lessons learned from a database analysis. Nat. Hazards 59/1, 285.

LEC, 2006. Lightning protection? Floating roof tank shunts, Lightning Eliminators & Consultants, Inc., In: Industrial Fire World, vol. 21, No. 6, November/December 2006, College Station, TX.

Maheshwari, B.K., Sharma, M.L., Narayan, J.P., 2005. Structural damages on the coast of Tamil Nadu due to tsunami caused by December 26, 2004 Sumatra earthquake. ISET J. Earthquake Technol. 42 (2–3), 63.

Necci, A., Antonioni, G., Cozzani, V., Krausmann, E., Borghetti, A., Nucci, C.A., 2013. A model for process equipment damage probability assessment due to lightning. Reliab. Eng. Syst. Saf. 115, 91.

Necci, A., Antonioni, G., Cozzani, V., Krausmann, E., Borghetti, A., Nucci, C.A., 2014. Assessment of lightning impact frequency for process equipment. Reliab. Eng. Syst. Saf. 130, 30.

NTHMP, 2001. Designing for Tsunamis—Seven Principles for Planning and Designing for Tsunami Hazards. National Tsunami Hazard Mitigation Program, USA.

NTHMP, 2013. NWS TsunamiReady® Program. http://www.tsunamiready.noaa.gov/

NTSB, 1996. Evaluation of pipeline failures during flooding and of oil spill response actions, San Jacinto River near Houston, Texas, October 1994. Pipeline Special Investigation Report, PB96-917004, NTSB/SIR-96/04, US National Transportation Safety Board.

OECD, 2012. Natech Risk Management—Natural Hazards Triggering Technological Accidents, Workshop Proceedings, Dresden, Germany, May 23–25.

Okubayashi, T., Suzuki, H., Kato, N., Yoshiki, N., Onishi, H., Takeuchi, K., 2016. A study of the method to reduce damages by tsunami applying flexible pipes. International Symposium on Natural and Technological Risk Reduction at Large Industrial Parks (NATECH 2016). Osaka, Japan, January 12–13. .

PIP, 2005. PIP STE03020 Guidelines for tank foundation designs, Process Industry Practices Structural, Construction Industry Institute, University of Texas, Austin, Texas, USA.

Rasmussen, K., 1995. Natural events and accidents with hazardous materials. J. Hazard. Mater. 40 (1), 43.

Renni, E., Krausmann, E., Cozzani, V., 2010. Industrial accidents triggered by lightning. J. Hazard. Mater. 184, 42.

Sakai, F., Isoe, A., Hirakawa, H., Mentani, Y., 1990. Seismic study on anchored cylindrical liquid storage tanks by static tilt tests using a large model of exact similitude (2nd Report). J. High Pressure Inst. Japan 28 (1), 30, (in Japanese).

Salzano, E., Iervolino, I., Fabbrocino, G., 2003. Seismic risk of atmospheric storage tanks in the framework of quantitative risk analysis. J. Loss Prev. Process Ind. 16, 403.

Santander, I., 2010. 350 mil personas del sur viven en zonas inundables, Coatzacoalcos, Agencia Imagen del Golfo, February 6.

Sierra, J., 2005. Sesenta municipios valencianos permiten que se construya en zonas inundables, El Mercantil Valenciano.

Steinberg, L.J., 2004. Natechs in the US: experience, safeguards, and gaps. In: Vetere-Arrellano, A.L., Cruz, A.M., Nordvik, J., P., Pisano, F. (Eds.), Proceedings of the Analysis of Natech (Natural Hazard Triggering Technological Disasters) Disaster Management Workshop, Ispra, Italy, 20–21 October 2003, EUR 21054 EN, European Communities.

Steinberg, L., Cruz, A.M., 2004. When natural and technological disasters collide: lessons from the Turkey earthquake of August 17, 1999. Nat. Hazards Rev. 5 (3), 121.

STREST, 2014. Report on lessons learned from recent catastrophic events. In: Krausmann, E. (Ed.), FP7 Project STREST—Harmonized Approach to Stress Tests for Critical Infrastructures Against natural Hazards, Deliverable 2.3.

Tar, T., Kato, N., Suzuki, H., Okubayashi, T., 2016. Experimental and numerical study on the reduction of tsunami flow using multiple flexible pipes. International Symposium on Natural and Technological Risk Reduction at Large Industrial Parks (NATECH 2016). Osaka, Japan, January 12–13. .

USACE, 1997. Flood Proofing—Techniques, Programs and References. US Army Corps of Engineers, Washington, DC.

USGS, 2003. Rupture in South-Central Alaska—The Denali Fault Earthquake of 2002, USGS Fact Sheet 014-03. United States Geological Survey, Reston, Virginia.

Whittaker, A., Moehle, J., Higashino, M., 1998. Evolution of seismic building design practice in Japan. Struct. Design Tall Build. 7, 93.

Yeh, H., Robertson, I., Preuss, J., 2005. Development of Design Guidelines for Structures That Serve as Tsunami Vertical Evacuation Sites, Open File Report 2005-4. Division of Geology and Earth Resources, Washington State Department of Natural Resources, Olympia, Washington.

Yokel, F.Y., Mathey, R.G., 1992. Earthquake Resistant Construction of Gas and Liquid Fuel Pipeline Systems Serving, or Regulated by, the Federal Government. US Department of Commerce, National Institute of Standards and Technology, Gaithersburg, MD.

Zama, S., Nishi, H., Hatayama, K., Yamada, M., Yoshihara, H., Ogawa, Y., 2012. On damage of oil storage tanks due to the 2011 off the Pacific Coast of Tohoku Earthquake (Mw 9.0), Japan. In: Proceedings of the Fifteenth World Conference on Earthquake Engineering, Lisbon, Portugal, September 24–28.

Reducing Natech Risk: Organizational Measures

E. Krausmann*, A.M. Cruz**, E. Salzano†
*European Commission, Joint Research Centre, Ispra, Italy; **Disaster Prevention Research Institute, Kyoto University, Kyoto, Japan; †Department of Civil, Chemical, Environmental, and Materials Engineering, University of Bologna, Bologna, Italy

Dealing with Natech risk effectively involves a wide range of prevention and mitigation measures that can be physical or administrative in nature. A mix of both types of measures is commonly required for optimum protection. This chapter introduces selected organizational measures for Natech risk reduction.

14.1 ORGANIZATIONAL RISK-REDUCTION MEASURES

Natech risk reduction involves structural but also organizational measures. In contrast to structural measures, which use engineered physical solutions, such as safety valves or containment dikes, to achieve protection goals, organizational measures are administrative programs and controls implemented to reduce risks. Organizational protection measures, often also called nonstructural measures, include educational and awareness campaigns, staff training, the establishment of safety practices and procedures including the monitoring of safety performance, and policies and laws. Considering that hazards can never be entirely eliminated from a hazardous installation using only technical protection measures, organizational control is needed to support accident prevention and mitigation (Saari, 2016). In fact, a lack of or bad organizational risk-reduction measures and practices, such as the absence of oversight mechanisms or bad management of change have caused or contributed to chemical accidents (Arocena et al., 2008).

14.2 NATECH RISK GOVERNANCE

Risks in general, and those stemming from existing or new technologies in particular, need to be properly governed to allow the society to benefit from these technologies while at the same time minimizing the potentially associated negative consequences. More specifically, risk governance involves all processes of interaction and decision

Natech Risk Assessment and Management. http://dx.doi.org/10.1016/B978-0-12-803807-9.00014-0

making among all actors that have a stake in a given risk, with the aim to identify, assess, manage, and communicate the risk. Clearly, good practice should be applied to risk governance for all types of risks.

From a Natech point of view, risk governance is becoming exceedingly important in modern times considering the increasing interconnectedness of society and industrialization, and the pace of new technological developments coupled with emerging hazards, such as climate change. There is concern among all stakeholders, such as government authorities, industry and civil society that risk-governance mechanisms might lag behind the processes that drive change in today's world, and that it might not be possible to effectively deal with new risks (Renn and Walker, 2008).

Since natural hazards may impact large areas simultaneously, addressing Natech risk requires an integrated risk governance approach to tackle both the safety of individual installations but also the potential interactions with other installations, lifelines, and nearby communities before, during, and after a natural-hazard event. The interdependencies of these systems may result in cascading events that can have short-, medium-, and long-term health, environmental, economic, and social impacts beyond the disaster areas. The Great East Japan earthquake and the Thai floods in 2011 highlighted the need to better understand infrastructure failure interdependencies and their risk governance. This means that the management of Natech risks requires incorporating parameters of the physical environment, such as lifelines, industrial facilities, and building stock, as well as organizational, social, and systemic factors into the analysis of natural-hazard risks (Cruz, 2012; Cruz et al., 2015).

Thus, the need to address Natech risk reduction as a territorial risk-governance issue is of the utmost importance. Natech risk reduction cannot be tackled as a problem of an individual facility, but only through a comprehensive and integrated risk-governance approach that involves all stakeholders.

The work of the International Risk Governance Council (IRGC) aims to support the better understanding and management of emerging global risks by developing concepts for risk governance, anticipating major risk issues, and providing recommendations on risk governance to key decision makers (www.irgc.org). In this context, the IRGC proposed an innovative risk-governance framework in an attempt to provide guidance on how to investigate, communicate, and manage particular risks (IRGC, 2012). This framework supports a comprehensive and integrated view of risk governance and comprises the following five elements:

- Risk preassessment: early warning and "framing" of the risk to provide a structured definition of the problem, of how it is framed by different stakeholders, and of how it may best be handled.
- Risk appraisal: combining a scientific risk assessment (of the hazard and its probability) with a systematic concern assessment (of public concerns and perceptions) to provide the knowledge base for subsequent decisions.
- Characterization and evaluation: using the scientific data and a thorough understanding of the societal values affected by the risk to determine if the risk is acceptable, tolerable (requiring mitigation), or intolerable (unacceptable).

- Risk management: actions and remedies needed to avoid, reduce, transfer, or retain the risk.
- Risk communication: how stakeholders and civil society understand the risk and participate in the risk-governance process.

The second and third element are very similar to the process for (Natech) risk assessment introduced in Chapter 7. The IRGC framework, however, does not only consider scientific evidence in the assessment but also includes risk perceptions, social concerns, and societal values. The IRGC also analyzed contributing factors that provide fertile ground for the emergence of new or the aggravation of existing risks, such as scientific unknowns, technological advances, perverse incentives, or a loss of safety margins (IRGC, 2010). The associated guidelines for emerging risk governance have been published recently (IRGC, 2015).

A comprehensive treatment of risk governance under conditions of increasing complexity is provided by Fra Paleo (2015) in which light is shed on the underlying structural factors, processes, players, and interactions which influence decision making, thereby either increasing or reducing disaster risks, including those of Natech accidents.

14.3 PREVENTION AND MITIGATION

A comprehensive risk assessment or lessons learned from past accidents and near misses can identify technical and organizational failures that may occur and drive the development and implementation of appropriate prevention and mitigation measures. With respect to learning from past events, Argyris and Schön (1974, 1978) contend that individuals and organizations involved in safety management should employ double-loop learning that allows them to detect and correct an error while at the same time critically examining and changing the values, assumptions, and objectives that might have led to actions with unwanted consequences in the first place.

Past experience showed that structural prevention and mitigation measures need to be supplemented by organizational measures at all actor levels to ensure the effective reduction of the risks associated with natural-event impacts at hazardous installations. Most importantly, industry operators should establish and promote a corporate safety culture that is reflected in a corporate safety policy or safety management system (OECD, 2003). This must include the periodic monitoring and/or reviewing of the safety performance of a hazardous installation, including the consideration of information related to natural-hazard risks (OECD, 2015).

Generally, during the design and construction stages of an industrial installation it is ascertained that the risks from the hazardous substances and processes present on-site are minimized. This includes the application of state-of-the-art design standards and codes of practice, which has to consider the risks from natural-hazard impacts where applicable. For earthquakes, for instance, seismic building codes based on a realistic assessment of the expected earthquake severity and the resultant loading on structures need to be implemented. Seismic design should be extended to also cover industrial equipment where not mandatory because the continued functionality of equipment containing hazmats and of safety-relevant auxiliary infrastructures is key to preventing a Natech accident (Krausmann et al., 2011a). It is essential that compliance with these codes is monitored. Since natural-hazard risks can vary over time and some industrial installations or infrastructures have a long operational life, natural hazards should not only be considered during the design stage but also during plant operation.

The best approach to preventing Natech accidents is naturally to keep hazardous installations away from natural-hazard prone areas via appropriate land-use-planning (LUP) arrangements and controls. LUP should consider the risks posed by natural events when considering the siting for hazardous industry, including the potential changes to the natural-hazard risk due to climate change. Authorities can, for instance, determine that certain areas, such as flood zones, may not be suitable for the siting of activities involving hazmats. Alternatively, they can call for additional protection measures or impose more stringent design, construction, and operational requirements in natural-hazard zones (OECD, 2015).

Once a decision on the siting of a new installation has been taken, the choice of the site layout with respect to the location of hazardous substances and processes can contribute significantly to reducing the likelihood of a Natech accident. For instance, if the siting of a new hazardous facility in a flood area cannot be avoided, it should be attempted to place equipment containing hazmats and other safety-critical plant components outside the projected inundation zone. Shut-off valves, for example, should be located above the predicted inundation levels as otherwise they might not be reachable during flood conditions. These risk zones may, however, not be static in time and a reassessment of these zones should be undertaken periodically to take account of newly available information or possibly changed boundary conditions related to natural-hazard frequency and severity. If these reviews show that the risks of Natech accidents have significantly increased over time at existing installations, the safety report should be updated, and retrofitting to comply with safety goals is advisable.

In natural-hazard prone areas, Natech-specific protection measures and systems should already be considered during the design stage of a hazardous facility. These measures might be mostly structural in nature. However, they need to be accompanied by procedures to make sure that plant personnel takes the correct actions in case of early warning, or during and after abnormal operating conditions, such as those caused by heavy rain, storms, earthquakes, etc. Examples of such procedures are the emergency shutdown of highly hazardous processes in case of power loss, for example, due to a lightning storm, or the careful monitoring of all processes during a plant's start-up to detect possible safety issues after a storm as early as possible. Another procedure is the deinventorying of tanks or pipeline systems exposed to natural hazards to reduce the hazardous materials at risk of being released in case of an accident. During Hurricanes Katrina and Rita in 2005, a significant number of pipeline breaks occurred in the Gulf of Mexico, which is testimony to the vulnerability of this type of infrastructure to natural-hazard impact. The much lower number of releases from these pipelines was attributed to the deinventorying of the pipelines in preparation for the storms (Cruz and Krausmann, 2009).

In areas subject to multihazard natural risks, for example, earthquakes followed by a tsunami, the consecutive impact of both natural events needs to be considered in the safety management of the hazardous plant. An earthquake preceding a tsunami could weaken or damage the facility, which would then be more vulnerable to the impacting tsunami wave. This is applicable to both shore protection systems and industrial facilities.

Considering that Natech risks are often underestimated or little understood, training and education of all actors involved in the reduction of Natech risks should be expedited. This holds in particular for plant personnel to ascertain that they are competent to carry out their tasks under normal, abnormal, and emergency conditions, but also for authorities to help them better evaluate the Natech risk and to support informed decision making. In addition, a dialog between all stakeholders should be facilitated to avoid the fragmentation of knowledge across different actors.

14.4 EMERGENCY-RESPONSE PLANNING

Emergency-response planning is at the interface between accident prevention and consequence mitigation and ensures adequate preparedness in case of an emergency. The control of Natech accidents requires special planning in terms of emergency management because major natural events, such as strong earthquakes or severe floods, may impact large areas affecting people, the building stock, as well as industry and other infrastructures. Natural hazards will most likely also impact safety measures, as well as directly affect emergency-response capacity, particularly if natural events have not been adequately factored into an installation's design and safety-management plan. Moreover, a natural disaster can contribute to the escalation of a chemical accident due to cascading events and interdependencies, often resulting in more severe consequences and complicating emergency response.

It is therefore obvious that emergency plans for accidents involving hazardous materials should take natural-hazard risks into account. However, emergency-response plans in the industry are typically developed for single accidents that are expected to occur during normal day-to-day plant operation, and seldom include the possibility of multiple releases that are common during Natech events. In addition, onsite emergency-response plans usually rely on the availability of external lifelines for accident mitigation which are often destroyed by the natural event. It is recommended that plant-internal emergency plans for mitigating hazmat releases during natural disasters should assume that off-site response resources are unavailable. Instead, they should provide for backup lifelines or specific emergency procedures to cope with the consequences of a Natech accident. In this context and to be conservative, safety barriers should be considered as absent or nonfunctional. Onsite emergency plans should also foresee means for the adequate control of ignition sources in the wake of hazmat releases to allow the safe use of emergency-response equipment, such as power generators, foam sprays, and pumps.

Off-site emergency-response plans for hazardous industry in natural-hazard prone areas need to consider the eventuality of hazmat releases from natural-hazard impact, and the effect of these releases, including fires and explosions, on the population and on rescue operations. These external response plans should incorporate the emergency evacuation of residents in the vicinity of the hazardous facility which might be challenging if access roads are blocked by debris, flooded, or destroyed. Although usually not considered, attention should be given to possibly violent reactions of released chemicals with floodwaters and the formation of secondary toxic or flammable vapors from possibly innocuous precursor chemicals (Cozzani et al., 2010).

Natural events can also damage response capabilities by affecting power and water supplies, access routes, and communication systems, rendering emergency response a big challenge. Several past Natech events (cf. Sections 2.2 and 2.3) showed that the hazmat releases may hamper emergency response to the natural-disaster victims by forcing first responders to abandon the area to not endanger their lives. An assessment of the vulnerability of the emergency-response resources is also called for in the context of Natech risk reduction.

In case of off-site consequences, local hospitals and clinics might have to treat people for toxic effects or burns. During natural disasters, it is, however, likely that hospitals might be overwhelmed by the onrush of natural-disaster victims, and therefore have only limited human and medical resources to deal with hazmat-exposure symptoms. In order to prepare for the eventuality of a Natech accident, local medical services should be informed about the risks at industrial facilities and make certain they have sufficient and suitable medication in stock for treating hazmat-release victims. Similar to medical personnel, local security forces play an important role during emergencies involving hazardous materials by informing and assisting the public during the evacuation, and by securing evacuation zones. This means that they are also at risk of exposure to toxic releases, fires, or explosions. It is essential that these units receive adequate training to better protect the population but also themselves (Girgin, 2011).

Generally, emergency-response plans drawn up at plant and community level should be periodically reviewed and tested to ensure that they address the potential consequences of natural-hazard impacts. This should be done proactively to avoid surprises during an emergency. The planning should also consider possible changes to the frequency and severity of some natural hazards due to climate change. Postaccident reviews serve to critically assess the performance of emergency response and offer an opportunity to improve response systems. Following the major Natech accidents during the Kocaeli earthquake in 1999, plant-internal accident investigations found that response resources were wanting during the emergency, and response capacities were subsequently increased. This included the installation of more and higher-capacity fire-fighting equipment, the improvement of the interoperability between plant-internal and off-site fire-fighting resources, and the inclusion of Natech scenarios in the updated emergency-response plan (Girgin, 2011).

It should be noted that careful consideration of conflicting emergency-management objectives such as the need to carry out search and rescue activities, while at the same time being forced to evacuate the same area because of a hazardous-materials release threat, is called for. This is aggravated by the fact that shelter-in-place to protect residents from the releases may not be feasible due to a loss of structural integrity of buildings by damage from earthquakes or other natural hazards. Steinberg and Cruz (2004) also found that panic flight behavior may be expected at facilities that house hazardous materials which suffered heavy damage. This finding indicates that mechanisms to deal with the lack of personnel at industrial facilities to handle emergencies involving hazardous materials following a large earthquake need to be identified.

The OECD (2015) suggests to integrate emergency planning for hazardous installations with emergency planning for natural disasters and civil defence, considering that these activities involve many of the same requirements. This would result in better coordinated and consistent emergency plans, as well as a coordinated command structure in case of an emergency.

14.5 EARLY WARNING

Early-warning systems play a pivotal role in the reduction of risks related to natural hazards, however, early warning is usually not available or practicable for mitigating Natech risks. Warning times for some natural hazards are often very short and might prove to be insufficient for taking preventive action at hazardous installations. Salzano et al. (2009) studied the conditions necessary for early warning for Natech risks and report that the effectiveness of such early-warning systems is characterized by the ratio of the available warning time, and the time necessary for implementing the required preventive action. The latter strongly depends on the type of equipment at risk, its operating conditions, the hazardous substances it contains, and the associated processes and actions of people and systems.

For early warning to be successful, the facility operator has to receive timely warning from a reliable source, e.g., from authorities, quickly evaluate this information, and act upon it. Table 14.1 gives an overview of the effectiveness of Natech early warning using the ratio of warning and action time as a basis.

Table 14.1 *Effectiveness of Natech-Specific Early-Warning Systems Based on the Warning Time t_{warn} and Action Time t_{act}*

t_{warn}/t_{act}	Characteristics	Effectiveness
$\ll 1$	Short warning time or slow preventive action	*Low*: Little time to implement preventive action
≈ 1	Warning time similar to time needed for preventive action	*Medium*: Some preventive action possible prior to natural-event impact
$\gg 1$	Long warning time or fast preventive action	*High*: Sufficient time for preventive action even if time-consuming

Early warning to prevent *earthquake*-triggered Natech accidents is the most unfavorable situation as warning times range from only a few seconds to a few minutes, depending on the distance of the hazardous installation from the earthquake's epicenter. The subsequent actuation of protection measures and systems would have to be extremely quick and rely on automatic processes, as human intervention would likely be too slow. Signals from seismic sensor networks installed onsite could activate safety interlock systems and valve closure on hazardous equipment or trigger emergency shutdown of dangerous processes (Salzano et al., 2009; Krausmann et al., 2011b). Valve closure is, however, not an instantaneous process, and van den Bosch and Weterings (1997) indicate that safety valve isolation for equipment at atmospheric pressure (e.g., hydrocarbon storage tanks) will take about 10 min while pressurized equipment can be isolated in 3 min. Early warning for earthquakes might therefore not prove to be very effective in preventing hazardous-materials releases and earthquake-resistant design should be prioritized.

River floods have warning times that can typically range from hours to days, thereby providing ample opportunity to mitigate the associated Natech risk. Possible risk-reduction measures under these conditions can include complete plant shutdown, depressurization of equipment, deinventorying of critical units, and the transfer of hazardous substances from inundation zones onsite to safer locations. Since these measures can be rather costly, reliable early warning and the minimization of false alarms are a prerequisite for the acceptability of these systems for controlling the Natech risk (Krausmann et al., 2011b).

The warning lead time for *tsunamis* depends on whether they are generated in the near or far field. If there is sufficient time, prevention actions like for floods can be taken for securing the installation. However, tsunamis can trigger Natech accidents not only by impacting process and storage units, but also by causing damage at connected port terminals and their loading and unloading infrastructure. In case of a tsunami warning, tankers moored at a refinery's oil terminal would require a warning lead time of a few hours to stop the product transfer, safely disconnect the loading arms, and move into deep waters to reduce the risk of a major oil spill (Eskijian, 2006).

Bouquegneau (2007) indicates that early warning is also possible for *lightning*-related hazards. Access to information from meteorological lightning-location systems can provide advance warning to operators, allowing them to take appropriate measures to disconnect sensitive equipment, stop hazardous processes, and protect open-air workers onsite.

References

Argyris, C., Schön, D., 1974. Theory in Practice: Increasing Professional Effectiveness. Jossey-Bass, San Francisco.

Argyris, C., Schön, D., 1978. Organizational Learning: A Theory of Action Perspective. Addison Wesley, Massachusetts, Reading..

Arocena, P., Núñez, I., Villanueva, M., 2008. The impact of prevention measures and organizational factors on occupational injuries. Saf. Sci. 46 (9), 1369.

Bouquegneau, C., 2007. Lightning protection of oil and gas industrial plants. In: Proceedings IX International Symposium on Lightning Protection, Foz do Iguaçu, Brazil, November 26–30.

Cozzani, V., Campedel, M., Renni, E., Krausmann, E., 2010. Industrial accidents triggered by flood events: analysis of past accidents. J. Hazard. Mater. 175, 501.

Cruz, A.M., 2012. Challenges in Natech risk reduction—Desafíos en la reducción de riesgos Natech. Revista de Ingenieria 37, 79.

Cruz, A.M., Kajitani, Y., Tatano, H., 2015. Natech disaster risk reduction: can integrated risk governance help? In: Fra Paleo, U. (Ed.), Risk Governance—The Articulation of Hazard, Politics and Ecology. Springer, The Netherlands.

Cruz, A.M., Krausmann, E., 2009. Hazardous-materials releases from offshore oil and gas facilities and emergency response following Hurricanes Katrina and Rita. J. Loss Prev. Process Ind. 22, 59.

Eskijian, M.L., 2006. Mitigation of seismic and meteorological hazards to marine oil terminals and other pier and wharf structures in California. Nat. Hazards 39, 343.

Fra Paleo, U., 2015. Risk Governance—The Articulation of Hazard, Politics and Ecology. Springer, The Netherlands.

Girgin, S., 2011. The Natech events during the 17 August 1999 Kocaeli earthquake: aftermath and lessons learned. Nat. Hazards Earth Syst. Sci. 11, 1129.

IRGC, 2010. The Emergence of Risks: Contributing Factors. International Risk Governance Council, Geneva.

IRGC, 2012. An Introduction to the IRGC Risk Governance Framework. International Risk Governance Council, Geneva.

IRGC, 2015. IRGC Guidelines for Emerging Risk Governance—Guidance for the Governance of Unfamiliar Risks. International Risk Governance Council, Lausanne.

Krausmann, E., Renni, E., Campedel, M., Cozzani, V., 2011a. Industrial accidents triggered by earthquakes, floods and lightning: lessons learned from a database analysis. Nat. Hazards 59 (1), 285.

Krausmann, E., Cozzani, V., Salzano, E., Renni, E., 2011b. Industrial accidents triggered by natural hazards: an emerging risk issue. Nat. Hazards Earth Syst. Sci. 11, 921.

OECD, 2003. OECD Guiding Principles for Chemical Accident Prevention, Preparedness and Response (second ed.), Series on Chemical Accidents No. 10, Organization for Economic Co-operation and Development, Paris.

OECD, 2015. Addendum Number 2 to the OECD Guiding Principles for Chemical Accident Prevention, Preparedness and Response (second ed.) to Address Natural Hazards Triggering Technological Accidents (Natechs), Series on Chemical Accidents No. 27, Organization for Economic Co-operation and Development, Paris.

Renn, O., Walker, K.D. (Eds.), 2008. Global Risk Governance—Concept and Practice Using the IRGC Framework. Springer, The Netherlands.

Saari, J., 2016. Accident prevention. Encyclopedia of Occupational Health and Safety. fourth ed. International Labour Organization, Geneva, Switzerland, (Chapter 56)..

Salzano, E., Garcia Agreda, A., Di Carluccio, B., Fabbrocino, G., 2009. Risk assessment an early warning systems for industrial facilities in seismic zones. Reliab. Eng. Syst. Saf. 94, 1577.

Steinberg, L.J., Cruz, A.M., 2004. When natural and technological disasters collide: lessons from the Turkey earthquake of August 17, 1999. Nat. Hazards Rev. 5 (3), 121.

van den Bosch, C.J.H., Weterings, R.A.P.M., 1997. Methods for the calculation of physical effects. Committee for the Prevention of Disasters, The Hague, The Netherlands.

Recommendations and Outlook

E. Krausmann*, A.M. Cruz**, E. Salzano[†]

*European Commission, Joint Research Centre, Ispra, Italy; **Disaster Prevention Research Institute, Kyoto University, Kyoto, Japan; [†]Department of Civil, Chemical, Environmental, and Materials Engineering, University of Bologna, Bologna, Italy

Many hazardous industrial activities provide society with indispensable goods and services. Some of these activities are considered particularly critical, such as refining, oil and gas transport and distribution, or the production of rare specialty chemicals, due to their criticality for ensuring human wellbeing and the smooth functioning of society.

Past Natech accidents highlighted the vulnerability of these industrial activities to natural-hazard impacts with consequences ranging from health impacts and environmental degradation to significant economic losses at local or regional level from asset damage and the associated business downtime. For major accidents, ripple effects on the economy can reach global proportions, resulting in a shortage of raw materials or intermediate products in the manufacturing industry, and causing global price hikes. Some of these accidents have also drawn attention to the increased risk of cascading effects during natural disasters and the challenges faced by emergency responders in combating accident consequences when lifelines have been downed by the natural event.

Unfortunately, Natech risk is bound to increase in the future. On the one hand, climate change is already affecting the severity and frequency of hydro-meteorological hazards, such as floods, heavy precipitation, or storms, while worldwide industrialization increases the number of technological hazards. On the other hand, human exposure and vulnerability is also growing with a trend toward increasing urbanization, and industry and community encroachment on areas that are natural-hazard prone.

Awareness of Natech risks is increasing and first attempts at systematically assessing and controlling this kind of risk are being made. However, the move toward a safer and more resilient society relies on the closing of a number of remaining research and policy gaps related to Natech risk reduction which require the attention of regulatory bodies, industry, and the research community.

Further awareness-raising efforts are needed to help stakeholders recognize the vulnerability of hazardous installations with respect to natural-hazard impact. This includes the recognition that vulnerabilities may also be tied to the nonavailability of protection measures, and of internal or external lifelines, which are also prone to

Natech Risk Assessment and Management. http://dx.doi.org/10.1016/B978-0-12-803807-9.00015-2

damage or failure during natural hazards. In this context it is important to note that interdependences between lifelines are not routinely assessed.

Considering the distribution of knowledge and competences across different actors, for example, industry, ministries in charge of civil protection, environment, or labor, it should be ensured that communication pertaining to Natech risks flows freely and effectively between these actors. Otherwise there is the danger of underestimating Natech risks with repercussions on safety legislation, technical standards and codes, and risk mitigation. Risk communication should therefore be improved in industry, and at all levels of government, be it national, regional, or local.

In some countries and for some industrial activities, current legislative frameworks explicitly address the risk of natural-hazard impacts on hazardous installations and can—in principle—provide a significant contribution to industrial safety. It is, however, crucial that competent authorities ensure the enforcement of these regulations for them to be effective in reducing Natech risks. Guidance on how to achieve the goals set out in the legislative framework should be developed to help industry comply with the legal requirements and to support authorities in evaluating if industry has met the associated safety objectives. Where missing, specific legislation for Natech risk reduction should be developed and implemented. Experience shows that in most situations risk mitigation worked best if required by law.

Risk assessment helps the operator to identify safety gaps in a hazardous installation and the associated risk-reduction priorities. For Natech risk assessment, no consolidated methodologies exist. Research should therefore focus with a priority on the development of methodologies and tools for Natech risk assessment and related guidance for industry. The accident scenarios identified in the risk assessment also support land-use- and emergency-planning decisions. In addition, an inventory of best practices for Natech risk reduction for different types of natural hazards should be developed and disseminated widely to combat the fragmentation of information potentially useful to all stakeholders.

In the assessment of Natech risks the impacts of climate change on natural-hazard severities and frequencies should be considered and appropriate action taken where additional vulnerabilities are identified. The adequacy of the design basis of hazardous installations and equipment against natural-hazard loading should be scrutinized in particular in light of climate change. This also includes an assessment of the adequacy of the protection measures in place. Natural-hazard maps that are kept updated and which include climate-change predictions can be helpful in addressing this problem.

Data availability is the bottleneck in Natech risk reduction. Only little data are available to researchers for learning lessons on the dynamics of Natech accidents and the effectiveness of prevention and mitigation measures. On the one hand, there is a tendency among the operators of hazardous installations to not voluntarily disclose information about near misses or accidents for fear of negative repercussions on their activity. Frequently, even company-internal documentation of accidents is missing. On the other hand, where reporting requirements exist, they usually apply only to those accidents whose consequences exceed a predefined severity threshold.

These two factors have caused a lot of precious data to be lost which translates into a missed opportunity to learn lessons from which the whole industrial-safety community could have benefitted. Consequently, the sharing of company data on Natech accidents and near misses should be promoted and facilitated by authorities. For effective information sharing to take place, industry might need assurances that the data will be used for future accident prevention and mitigation rather than to assign blame. Where required, accident data can be anonymized. It would also be an important step forward to separate the reporting criteria from consequence severity to ensure that also low-impact accidents or near misses are captured. These events have often proven to be equally important from a lesson-learning perspective.

With Natech risk reduction cutting across several disciplines, different stakeholder groups need to be trained to ascertain that they have the adequate knowledge and skills to carry out their tasks in situations that deviate from the normal operating conditions they have been trained for, e.g., in case of natural-hazard impacts, and to properly handle the possible complications that frequently arise during Natech accidents. These groups comprise not only personnel employed at hazardous installations, but also authorities in charge of chemical-accident prevention and civil protection.

Closure of the aforementioned gaps will require close collaboration between scientists, engineers, operators, and policy makers in an interdisciplinary effort to address Natech risk reduction in a comprehensive way that is as cost-effective as possible. Public–private partnerships could play an important role in linking science, practice, and policy and should be explored in this context.

The many Natech accidents in the wake of the Tohoku earthquake and tsunami in 2011 took the world by surprise, in particular because they happened in a country with high levels of preparedness and advanced emergency-response capacities. The situation is even more challenging in the developing world where basic industrial-safety knowledge is often lacking and which is therefore ill-equipped to address Natech risks effectively. The outlook is, however, promising as awareness of Natech risks has grown worldwide post-Tohoku, and scientists have started to join forces with industry and government in an effort to tackle the issue. As devastating as it was, by drawing the world's attention to important gaps in accident risk management, the Japan disaster has created an opportunity to make the world a safer place.

Glossary

Accident Unintended and unforeseen event or series of events and circumstances that results in one or more undesirable consequences.

Accident scenario Sequence of events leading to adverse consequences.

Atmospheric tank Vessel with a fixed or a floating roof in which hazardous substances are contained at atmospheric pressure. Atmospheric tanks are commonly used for the storage of large quantities of liquid hydrocarbons.

BLEVE Boiling liquid expanding vapor explosion. This is an explosion that is caused by the rupture of a vessel that contains a pressurized toxic or flammable liquid that is heated above its boiling point. If the liquid is flammable, a BLEVE is associated with a fireball in case of immediate ignition, or with a flash fire or a VCE if ignition is delayed.

Cascading event Chain of events in which a primary event triggers a secondary event, which in turn can cause a tertiary event, etc. During Natech accidents the risk of cascading events is generally higher than during conventional technological accidents.

Consequence Outcome of an event which can be positive or negative. An event can have more than one outcome.

Damage state Category of equipment damage as a function of natural-hazard severity.

Disaster Natural or man-made event resulting in widespread human, environmental, economic, or material losses. The adverse consequences of a disaster can exceed the ability of the affected community or society to cope using its own resources.

Domino event See Cascading event.

EC European Commission. The EC is the executive body of the European Union. As such, it is in charge of proposing legislation, implementing decisions, upholding the EU treaties, and managing the day-to-day business of the EU.

EU European Union. The EU is a supranational political and economic union which includes 28 independent and democratic Member States in the European continent.

Event tree Inductive analytical diagram in which Boolean logic is used to analyze the progression and final outcomes of an event. Event Tree Analysis is a process that traces forward in time or through a chain of causes (see also Fault Tree Analysis).

Explosion Sudden and violent release of energy accompanied by pressure and/or heat.

Fault Tree Analysis Deductive process using Boolean logic that aims to understand the causes or combinations of causes that lead to an undesired top event, usually loss of containment (see LOC). Fault Tree Analysis traces backward in time or through a causal chain.

Fireball Fire event originating from the immediate ignition of a flammable vapor or gas cloud caused by a prior catastrophic loss of containment (see BLEVE).

Flash Fire Sudden combustion of a mixture of flammable substance and air. Flash fires are characterized by high temperature, short duration, and a rapidly moving flame front. Flash fires do not cause blast waves (see BLEVE).

F–N curve Plot of the cumulative frequency (F) of different accident scenarios against the number of potential casualties (N) associated with these scenarios.

Fragility curve Probability of failure or damage as a function of the degree of natural-event loading experienced.

Frequency (relative) Likelihood of an event over a specific time interval, usually expressed as $year^{-1}$.

Natech Risk Assessment and Management. http://dx.doi.org/10.1016/B978-0-12-803807-9.00019-X

Hazard Source of danger. A hazard does not necessarily lead to harm but represents only a potential to result in harm.

Hazardous substance/material Biological, chemical, radiological, or nuclear agent which poses a threat to health, animals, or the environment.

Hazmat Hazardous material.

HAZOP Hazard and Operability Study. HAZOP is a structured and systematic process to analyze potential hazards to personnel or equipment by reviewing the design of a planned or existing process or operation to identify potential issues of concern. The advantage of HAZOP is that in addition to hazards it also addresses operability problems.

Individual risk Probability for an individual to suffer ill effects at a specific point around a hazardous installation per given time period.

Isorisk curve Level of equal individual risk around a hazardous installation.

Jet fire Fire resulting from the release of pressurized flammable materials from an aperture with significant momentum.

Lesson learned Knowledge gained from investigation, study, or other activities in regard to the technical, behavioral, cultural, management, or other factors, which led, could have led, or contributed to the occurrence of an accident or a natural disaster.

LNG Liquefied Natural Gas. LNG is a natural gas that for ease of transport and storage has been converted to liquid form, thereby significantly reducing its volume.

LOC Loss of containment. Event in which hazardous materials are released from the equipment they were contained in.

LPG Liquefied Petroleum Gas. LPG is a flammable hydrocarbon mixture and is also referred to as propane or butane.

LUP Land use planning. LUP helps ensure that hazardous installations are separated by appropriate distances from other installations and developments, thereby preventing negative consequences.

Major accident Major toxic release, fire, or explosion whose impacts on human health and the natural environment exceed specific severity thresholds.

Missile Projection of a fragment in the wake of equipment rupture. Missiles can cause damage to other equipment on- or off-site, thereby increasing the risk of cascading events.

Mitigation Actions or measures taken to reduce the impact of hazardous-materials releases if they occur. Risk mitigation, on the other hand, aims to lower the risk of an accident (see also Risk reduction).

MMI Modified Mercalli Intensity scale. This scale is used to measure the intensity of an earthquake. It is based on observed effects and ranges from a scale of I (not felt) to XII (total destruction).

Natech Technological accident involving the release of hazardous materials caused by a natural hazard. The releases can be chemical, biological, or radiological in nature.

NOAA National Oceanic and Atmospheric Administration (NOAA) of the US Department of Commerce. NOAA's mission is to understand and predict changes in climate, weather, oceans, and coasts, and conserve and manage coastal and marine ecosystems and resources.

OECD Organisation for Economic Co-operation and Development. The OECD promotes policies that will improve the economic and social well-being of people around the world.

Organizational measure Any measure not involving physical construction that uses knowledge, practice, or agreement to reduce risks and impacts, in particular through policies and laws, awareness raising, training, and education (see also Structural measure).

PDF Probability Density Function. A PDF of a continuous random variable is a function that describes the relative likelihood for this random variable to take on a given value.

PGA Peak ground acceleration. PGA is the maximum acceleration of the ground during earthquake shaking at a specific location. It is commonly expressed in percentage of the acceleration of free fall, g, or in m/s^2 (where $1\ g = 9.81\ m/s^2$).

PGV Peak ground velocity. PGV is the maximum velocity of the ground subjected to an earthquake. It is measured in m/s.

Pool fire Burning pool of liquid, more specifically a fire burning above a pool of evaporating liquid.

Preparedness Measures taken before an adverse event in order to be prepared to adequately respond when the event occurs.

Pressurized tank Vessel in which hazardous substances are stored or processed at pressures higher than atmospheric pressure.

Prevention Actions or measures taken to reduce the likelihood of damage and occurrence of a hazardous-materials release.

Probability Measure for the likelihood of a random event expressed as a number between 0 and 1.

Probit method Probability unit method. It relates the magnitude of an effect (dose) to the extent of damage caused (response). The probit variable can be easily converted into a probability or percentage.

PSHA Probabilistic Seismic Hazard Analysis. PSHA aims to quantify the exceedance rate of different ground-motion levels at a specific location considering all possible earthquakes.

PTHA Probabilistic Tsunami Hazard Analysis.

QRA Quantitative Risk Analysis or Assessment. Process to estimate the risk of an event quantitatively. The outcome of a QRA is usually expressed in terms of fatalities or economic losses.

Return period Recurrence interval of a natural event, that is, an estimate of the likelihood for a flood, earthquake, etc. to occur.

Risk Combination of the frequency, or probability, of occurrence and the consequence of a hazardous event. Risk therefore includes the likelihood of conversion of a hazard into actual delivery of injury, damage, or harm. Risk is always subject to uncertainty related to the occurrence of the event.

Risk analysis Process that identifies hazard sources and estimates the associated risk.

Risk assessment Risk analysis and risk evaluation.

Risk evaluation Process in which the estimated risk is compared to given risk acceptability criteria.

Risk management Process that includes risk assessment, risk treatment, risk acceptance, and risk communication.

Risk reduction Actions that aim at lowering the probability of an adverse event, limiting its consequences, or both. The associated measures can be structural and organizational.

Risk state Category for the intensity of loss of containment.

Risk treatment Selection and implementation of measures to modify the estimated risk.

Societal risk Risk to the community from identified risk sources. Societal risk is usually expressed as *F–N* curves.

Structural measure Any physical construction to reduce or avoid possible impacts of hazards, or application of engineering techniques to achieve hazard-resistance and resilience in structures or systems (see also Organizational measure).

VCE Vapor cloud explosion. A VCE results from the ignition of a large cloud of flammable vapor, gas, or mist in which significant overpressure is produced.

WGCA Working Group on Chemical Accidents of the OECD.

Index